BASIC CONCEPTS OF MODERN PHYSICS

# BASIC CONCEPTS OF
# MODERN PHYSICS

QUANTA, PARTICLES, RELATIVITY

## GEORG UNGER

Portal books ≈ 2024

2024
**Portal**books
An imprint of SteinerBooks/Anthroposophic Press, Inc.
834 Main Street, PO Box 358, Spencertown, NY 12165
www.steinerbooks.org

Copyright © 2024 by Portal Books, an imprint of SteinerBooks/Anthroposophic Press, Inc. Translated from *Grundbegriffe der modernen Physik Quanten, Teilchen, Relativität: Vom Bilden physikalischer Begriffe - Teil III* (Verlag Freies Geisteslaben, 1967). Originally published in English as *Forming Concepts in Physics* (Parker Courtney, Chestnut Ridge, NY, 1995). The text was revised for this edition. All rights reserved. No part of this publication may be reproduced, stored in a retrieval system, or transmitted, in any form or by any means, electronic, mechanical, photocopying, recording, or otherwise, without the prior written permission of the publisher.

Revised Second Edition

Translated by Hanna Edelglass

LIBRARY OF CONGRESS CONTROL NUMBER: 2023946954

ISBN: 978-1-938685-49-1

## Contents

| | |
|---|---|
| Preface | ix |
| Introduction | xi |
|    The Role of Thinking in Physical Research | xi |
|    "Relations and Laws" | xii |
|    The Worldview: | |
|       Does It Belong in a Discussion on Thinking in Physics? | xiv |
| **1. A New View of Nature?** | 1 |
|    Phenomenon and Natural Law | 1 |
|    What, then, Is a Natural Law? | 8 |
|    Natural Law | 12 |
|    Concepts and Laws Have Content through Thinking | 17 |
|    Cognitive Character in Concept Extension | 19 |
|    Parallel Postulate as an Example of | |
|       the Way an Axiom Is Experienced ("Actually Thought") | 20 |
|    The Creation of a Concept (Transfinite Numbers) | 22 |
|    Does a Picture of the World Exist | |
|       That Arises from Thinking That Grasps Reality? | 24 |
|    Matter | 26 |
|    Quantum Physics Requires a New Concept | 26 |
|    New Physics Requires Mathematically Ideal Elements | 27 |
| **2. The Method of Gaining Knowledge** | 31 |
|    The Role of Thinking Revisited | 31 |
|    Cognition and Thinking | 31 |
|    Natural-Scientific Method | 32 |
|    The Experience of Thinking | 34 |
| **3. The Transition to Twentieth-Century Physics** | 41 |
|    What Actually Happened | 41 |
|    Present Shape of the Atomistic Worldview | 43 |

4. **The New Phenomena**     49

   *The Limits of Classical Physics*     49
   *Phenomenal Atomism*     51
   *The Consequences of Phenomenological Atomism*     53
   *More about the New Phenomena*     59
   *A Concept of Matter and the Boundaries of Sensory Reality*     60
   *Matter as an Invariant*     60

5. **Foundational Concepts of Quantum Theory according to Blokhintsev**     65

   *Quantized Light*     65
   *Atomism*     68
   *Bohr's Theory*     69
   *De Broglie's Waves*     70
   *Phenomenological Results*     71
   *Statistical Interpretation*     72
   *The Waves of the New Mechanics:*
      *Structures of Information as Physical Reality*     74
   *The Laws of Motion of Particles*     76

6. **The Concept of Probability**     77

   *Historical Notes*     77
   *Comparison with Geometry*     78
   *Intuitive Basic Assumptions*     79
   *The Subjective Perspective*     80
   *Chance and Necessity*     83
   *The Elementary Concept of Probability*     84
   *Elementary Calculus of Probability*     85
   *Connection between the Logical Concept of the "Accidental"*
      *and the Mathematical Concept of Probability*     87
   *Forming the Concept of Probability*     88
   *Overcoming Determinism*     88
   *Mixing Characteristics*     89
   *The Law of Large Numbers*     90
   *Summary and Review*     92

7. **The Theory of Relativity and Its Conceptual Constructions**     97

   *Observations regarding the Speed of Light*     97
   *Kinematics as Free Creation of the Human Spirit*     99
   *The Theory of Relativity as a Non-Galilean Metric*     101
   *Critique of Simultaneity*     103

| | | |
|---|---|---|
| | Length Contraction and Time Dilation | 104 |
| | Metrics in General | 107 |
| | The Metrics of Velocities | 108 |
| | Relativistic Mechanics | 110 |
| | Mass-Energy Equivalence | 110 |
| | Speed of Light as a Limiting Speed | 112 |
| | Synopsis of the Special Theory of Relativity | 113 |
| | So-called General Relativity Theory | 114 |
| | The Observer in the Box | 115 |
| | The Tensor Calculus | 116 |
| | Geometrization of Mechanics | 117 |
| | Concluding Remarks | 118 |
| 8. | **Concrete Concept Formations** | 121 |
| | Classical Laws of Conservation and Matter | 121 |
| | Matter as State | 122 |
| | Energy | 123 |
| | The Concept of the Quantum Mechanical State | 125 |
| | Bohr's Model | 128 |
| | The Novelty of the World of Particles and Quantum Events | 129 |
| | Theses | 131 |
| | The Fundamental Probability Propositions of Quantum Physics | 132 |
| 9. | **The Phenomenology and Mathematics of the New Physics** | 135 |
| | Phenomenology in the New Physics | 135 |
| | Excursion into Concepts That Are Grasped… yet Not Grasped | 137 |
| | The Significance of Mathematical Concept Formations in Classical and Modern Physics | 138 |
| | Equations and Differential Equations | 140 |
| | The Differential at the Border of the Imperceptible | 142 |
| | The Path from Differential Equations to Operators | 143 |
| | Mathematical Structure as Substitute for Naïve Reality | 146 |
| | The Superposition of Quantum Mechanical States | 147 |
| | Resolution of a State into Component States | 149 |
| 10. | **Physical Worldview and Spiritual Science** | 153 |
| | The Relationship between Suprasensory Entities and Sensory Perception | 153 |
| | The Suprasensory World in the Sense of Anthroposophy | 155 |

Spiritual Entities Have Relationships That
  Cannot Be Derived Solely from Sensory Experience    157
The Distance between the Subsensory World of Elementary
  Particles and the Suprasensory World of Elemental Beings  162
Is the Current Path of Theoretical Physics
  the Only One Possible?    169
Total Gestalts in Physics    171
Energy-free Transmission of Information
  and the Future of Physical Formulas    173
Rudolf Steiner's Counterspace—A Thought Form
  That Still Needs to Be Made More Concrete:
  The Duality Principle    176
Counterspace    178
Possible Physical Applications    181
Nonrelativistic Simultaneities    183
Total Gestalts    183
Homeopathic Dilutions (Potencies)
  and Cosmic Influences as Examples    185
Summary    192

**Cited Works**    195

**Index of Names**    197

# Preface

The book *Quantum Reality* by Nick Herbert gives a concise account of the philosophies that wrestle with the consequences of quantum theory and its modern experimental confirmation. My approach is different in one decisive aspect: I hold that human thinking is *not limited* to the thinking subject. The ability to communicate ideas and concepts from one individual to another is not "subjectively based on the similarity of brain functions." Instead, I maintain—and try to make it clear in this book—that, when we form concepts appropriate to our experience, the content of these thoughts has to do, *essentially,* with what we think about. In other words, my thoughts are part of the very world that I want to know. Rudolf Steiner's contribution was to show that this thesis can be verified by epistemological introspection; this is what can give us confidence in cognition.

The author is grateful to Hanna Edelglass for her translation and to Stephen Edelglass and David Booth for their efforts in illuminating discussions concerning both the content and the style.

*Georg Unger*
*Sant'Abbondio, Switzerland*
*August 1995*

*Even today, all of our education, especially in the exact sciences, is saturated with the residues of representations that have been "superseded scientifically." In view of this situation, it seems appropriate to present a new picture of the world just as uncompromisingly as Giordano Bruno and Galileo did for science that was modern in their day.*

# Introduction

It would be a misunderstanding to think that this book aims to deny recognized results or overthrow important theories through its examination of *concept formation in physics*.

The objection could be made that from all the discussions that I hold with the reader in what follows there results not one new experiment, that the point in question is an "idle game with concepts" that are neither true nor false but are simply *irrelevant* according to the doctrine of *physical positivism* (see Pascual Jordan).

But this objection, too, would be a fundamental misunderstanding not only of what is presented here but also of the possibilities of human thought altogether. This is true because *physical positivism* justifies itself by the necessity of critically testing the simplest statements. However, it is then equally true that, on the one hand, too much might be thrown overboard by such testing while, on the other hand, not all the remnants of earlier natural-philosophic views will have been removed.

To rephrase this dilemma: The revolution of thinking in the twentieth century is not yet *radical* enough in many respects. At the same time, it restricts the view of new possibilities quite unnecessarily when, in its wake, the role of thinking in forming knowledge is confined too narrowly.

## The Role of Thinking in Physical Research

With some justification, physical positivism puts forth the condition "that every scientific pronouncement has real content and meaning only insofar as it expresses relationships and laws in the material of our experimental experience" (P. Jordan, p. 141).[16]

## "Relations and Laws"

In context, the emphasis of Jordan's statement lies in its latter part: Only insofar as the relationships and laws relate to *the material of our empirical experience* do the pronouncements expressed by them contain real content and meaning—for physics! From this point of view, it remains uncontested that the relationships and laws of pure mathematics have content and meaning—for mathematics! Most definitely being contested, however, and with a certain brutality, is that the pronouncements of physics can discern something as regards the "true essence of things."

One form of this, according to Gustav Mie, is that we want to get to know nature only to dominate it. Werner Heisenberg, with a much more careful formulation, puts it this way: "The natural scientist therefore has to avoid the direct merger of the basic concepts on which he rests his science with the world of the senses." This situation follows from the fact that with the "explanation of sense-perceptible qualities of matter from the behavior of atoms," it becomes clear "that sense-perceptible qualities cannot be ascribed at all to the final building blocks of substances in a simple way" (Heisenberg, p. 98).[10] However we look at it, science always strives for *relationships* and *laws*.

Here, the schizophrenia of modern humanity clearly emerges: Science does not engage in physics for the sake of ultimate answers. Instead, researchers try to *broaden* and *deepen* relationships and laws at the present boundaries of science that are not sufficiently clear—perhaps to *change* these relationships through the responses of nature to suitably posed experimental questions—to understand what changes are necessary to produce a more encompassing theory. By carrying out research in this way, the physicist's thinking is like that of any human being when faced with an external reality that

*Introduction*

initially appears as an enigma. They solve and *grasp* such a reality by penetrating it with their own thought. However, in theorizing about what they do, those physicists state something quite different—that their thinking is completely incapable of understanding the true nature of things. It allows them, they claim, only to gain starting points for their *formulas*, in more or less arbitrary pictures (the fashionable word is *models*) of whose limits of validity they are aware. These formulas then lead to results that can be translated back into measurable quantities.

Only a negligibly small part of the resulting measurements serves the control of nature in the narrow sense of the word—in technology, for instance. The overwhelming part of all measurements serves the confirmation or rejection of certain theoretical ideas, or at least the determination of constants in more or less secure theoretical systems. Measurements can also serve to estimate whether certain experiments, which seem reasonable in principle, can actually have results that offer hope of success. If one, therefore, takes the phrase "control of nature" in the sense offered by the research attitude of modern physicists themselves rather than by theorizing about it, then it becomes merely a self-consciously humble paraphrase of *knowledge!* Just that. Even if one, as Faust, wants to "envision all creative activity and seed," it is still nothing else than *mastery of nature through thinking.*

It is a common understatement that natural science strives for the *control* of nature rather than its *intellectual* mastery. It seems that science is somehow ashamed of what is going on in man while he is cognizing. This is a result of the supposed insight that thinking has no content that belongs to nature. It has only social communication value for describing actions (experiments) that have to be undertaken and for reporting their consequences (needle deflections, for example). I want to follow the consequences of a new description of the goal of knowledge in the next paragraphs.

To begin with, we can see that no problem arises for the materialist, but even materialists use *thinking* to articulate their views of "matter" or to give it dialectical foundations. They also use thinking when trying to come to terms with the results of the experiments of twentieth-century physics, which very much contradict the naïve concept of matter. To them, thinking is a material process, but only when they speak *about it*. When doing research, they behave just like the researcher that they call "idealistic" or "subjectivistic." The idealist and the materialist both believe in the *results* of their thinking that provide them with a certain image of the process. Those results are taken seriously. A difference arises when, afterward, one of them denies the power of thinking to make valid statements about the being of things, while the other uncritically takes the results of thinking for the existential basis of things. The first, paradoxically, is called the "idealist physicist"; the latter ascribes to thinking, believed to be a material process, the ability to take hold of matter. By that, the materialist's matter is something completely intangible. Its existence is merely the permanent basis for all physical changes. It is postulated.

## The Worldview: Does It Belong in a Discussion on Thinking in Physics?

This book has to contain research about the real form of certain physical concepts, whether in the usual theoretical fields or within more methodically conscious and therefore, in a certain sense, more legitimate proceedings. And it has to contain a chapter in which the author resolutely makes the attempt, with the concepts that *he* has won so far, to create a view of the world.

I have asked myself the question of whether such a chapter should be at the beginning or at the end. Didactic reasons speak for the second placement, for then conceptual constructions are analyzed and purified before they are used. However, the other procedure—to expound upon the physics of matter and particles along with the description of a *world picture*—reflects my intention to avoid sacrificing life for the appearance of something systematic. In this way,

*Introduction*

the general ideas take form even as the individual expositions are being formulated.

One reason to begin with the world picture may be added here. Physics, ever since the time of Galileo, has been working on the formation of a world picture. The physicists themselves only became conscious of this late within their factual work; this happened when they saw themselves in their own specialty facing obstacles and difficulties of understanding that were nothing less than the results of their own earlier (specialized) views that had been vulgarized into the popular worldview made absolute. Even today, all of our education, especially in the exact sciences, is saturated with the residues of representations that have been "superseded scientifically." In view of this situation, it seems appropriate to present a new picture of the world just as uncompromisingly as Giordano Bruno and Galileo did for science that was modern in their day.

I see one way to prevent the misapprehension that the scientific philosophizing that I am undertaking here might follow a fixed track: Formulate as clearly as possible the new kinds of consequences to which I see myself led by logic, doing so with honesty and with the courage of my convictions. This will at least dispel the accusation that I am striving to make a worldview *scientifically* acceptable that has actually been preordained by anthroposophy or that I am at carrying the contents of this "teaching" dogmatically into scientific thought.

*Only in the course of this investigation may we hope to uncover the true function of thinking in physics. In any case, in even the simplest parts of physical science many conceptual distinctions lie hidden, whether we speak of phenomena or effects. Who would deny that the rainbow is a single thing rather than merely the play of refraction and dispersion summed over countless drops?*

Chapter 1

# A New View of Nature?

## Phenomenon and Natural Law

Prescientific observations of nature are *connected* to the thinking, searching mind. Naïve people *see* the connection in the same realm of existence as the facts themselves; they make the connection objective—for example, making warmth into a pseudo-substance (*phlogiston*). Equally naïve, they assume that the objects of their surroundings are *real* and that they, taken together, form the real world, about which thinking can produce a more exact understanding. When people have succeeded in recognizing these observed connections as necessary consequences of the real elements and their effects, they have *explained* something. That was an ideal: *the explanation of nature;* tacitly, though perhaps under other names, it still is.

The best-known example is that of tracing back falling to a single common cause—that is, gravitation working between all bodies, even between the great Earth and the small bodies of daily experience. This example illustrates both the *power* of such a thought and its limits, which will become conscious to the critical scientist.

The power appears because now the movement of the Moon with respect to the Earth and the movement of the Earth and planets around the Sun will all obey the same *law*. As an aside, this statement also contains the clarification of the fact that, apart from interferences, all bodies fall at the same rate. Carl Friedrich von Weizsäcker (p. 107)[34] rejects the myth:

> Galileo had broken ground for science in that he had described the world as we really experience it.... Galileo took a great step in that he dared to describe the world as we do not experience it. He drew up laws that, in the form in which he pronounced them, could never be valid in real experience and therefore could never through any single observation be confirmed. But, in exchange for that, the laws he described were mathematically simple. Thus, he opened the way for mathematical analysis that divided the complexity of real experiences into single elements. Scientific experimentation is different from everyday experience in that it is led by a mathematical theory that poses a question and is capable of pointing to an answer.

Thus, Galileo contradicted everyday experience with the postulate that all bodies fall with the same acceleration. With the hypothesis that this is valid in empty space, he created the strong motivation that led his student Evangelista Torricelli to create a vacuum and thus confirm Galileo's prediction (p. 108).[34]

Using the same example, another aspect of this thinking can be illustrated: quantitative control of phenomena as a *touchstone for correctness*. One becomes conscious of the *limits* when one makes oneself conscious of the tacit *preconceptions* of the thought of gravity. To begin with, the universal gravitation between any two material bodies is at least as unexplained as the fact that at the same place all small bodies (small in comparison to Earth) accelerate (approximately) at the same rate. This observation, by the way, was later a motive for Einstein's theory of gravitation. Second, to explain falling at the same speed, the law of inertia is needed in order to understand that the heavier body is attracted more strongly by the earth, but balancing that, provides proportionally greater *inertial* resistance, and *therefore* does not fall faster than the lighter body.

Robert Lindsay and Henry Margenau severely oppose this naïve interpretation of physics as the explainer of the world:

## A New View of Nature?

> Physics has nothing to say about a possible real world lying behind experience.... One often hears the statement that the task of physics is to describe or explain the behavior of the material world. We cannot help feeling that this is meant to imply the existence of such a world, though what the adjective "material" here means is by no means clear. The physicist has been striving for years to attach a clear meaning to the term *matter,* and undoubtedly, we have reason to believe that the concept means much more to us today than to the physicists of fifty years ago. However, to have a clear understanding of a physical concept like matter is not at all equivalent to the assumption that there is a "real" world behind our sense perceptions, which is responsible for the existence of matter. (p. 2)[19]

We can easily make a connection with this clear, precisely phrased rejection of the naïve concept of the real world. Of course, I do not want to save the *reality* of the naïve realist, but I also don't want to abandon that feeling that these same authors observe as follows: There is perhaps no harm in such an assumption—in fact, certain minds may find that it enables them more firmly to grasp and feel confidence in physical theories; yet it must be stressed that the assumption is no necessary part of physics, and that in a logical development of the subject the safest course is to omit it entirely. It is possible, indeed, to take the view that adopting such an assumption as part of the physicist's stock-in-trade involves a handicap.

The authors Robert Lindsay and Henry Margenau feel that its only advantage is the conviction that we approach the assumed reality in small, sure steps with the possibility, however, that this belief could lead us astray by a dependence on some halfway successful theories that allow too small a margin for revision in the face of new experiences. "Thus, the term *physical world* should not be construed as being identical with real world."[19]

One can agree totally with Lindsay and Margenau's view that the *naïve* concept of a "real world" behind appearances can be an obstacle to progress of theory under certain circumstances.

Nevertheless, one does not immediately have to renounce the ideal that physics has something to do with *reality* or an aspect of reality that has been rendered prominent in a certain way. To begin with, let's allow the ordinary view of physics to stand. It is a field accessible through repeatable experiments, measurement, and quantitative treatment.

It is understood that not every single observation or arbitrary measurement, however exact it may be, presents us with a scientific fact. Remember our beginning: We see *connections,* naïvely at first, and we naïvely project them out into the world. Then there followed Lindsay and Margenau's criticism that, as far as the naïve consciousness is concerned, is entirely justified. It is proper to examine *what* it is, between whose elements the connections exist. Then we have to explain the nature of these elements satisfactorily, at least so far as to understand their relationship to that reality we are looking for. That *our* concept of reality cannot be that of naïve observation anymore is clear.

To have a word, let's first name the *what* with the hallowed expression *phenomenon*. One reason for this word is that physics in its early stages occupied itself with phenomena, not with *effects,* and because I also want to stress that we start with what *appears to us*. The phenomenon might be the rainbow or the fact that a magnet attracts pieces of iron.

An argument could be raised that our *phenomenon* is complex and unexplained. It becomes physically transparent only by being dissolved into its single effects.* All these constitute the complex phenomenon. Physics deals with the rainbow in such a way that, after penetrating these single facts, there remains only this: One observes colored light effects that—after reflection at a definite angle from raindrops—appear on a cone of sight having a specific

---

* In the case of the rainbow, for example, it belongs to the description of the phenomenon that the rainbow wanders with the observer; that it appears high when the sun is low; that it is visible only when the sun is behind the observer, while before the observer is a veil of rain; and that it appears sometimes as a single bow and other times as multiple bows.

vertex angle, the axis of which points backward to the sun. This explanation comprehends at once the position of the bow and its relation to the sun and also to the movement of the observer. All that matters is that the single drops must be in position to shine varicolored, *in a regular sequence of color at a specific angle* between the sun and the observer. This all becomes quantitatively describable using the *general* facts of refraction, dispersion, and interference for some details, so that the phenomenon is *explained*. In other words, the phenomenon has been incorporated into a theory and with that, finally, is an inessential, specific, although striking, and speaking unscientifically indeed, a *beautiful* phenomenon. The original phenomenon may be complicated. Here lies an enigma! Admittedly, it does not consist of sense-perceptible elements that can be seen at once; however, the theory requires many concepts for its statement alone. (I will have to return this when we better understand the role of thinking.)

Only in the course of this investigation may we hope to uncover the true function of thinking in physics. In any case, in even the simplest parts of physical science many conceptual distinctions lie hidden, whether we speak of *phenomena* or *effects*. Who would deny that the rainbow is a single thing rather than merely the play of refraction and dispersion summed over countless drops?

We bring order into the apparent chaos of sense perceptions by *thinking*. Let's not overlook our own thoughts, otherwise we pay for it as in the example of an imagined world behind the senses, a *real world,* our own notion that remains with us only when we have forgotten *our* own thinking activity.

When we saturate, or enmesh, the sensory world in thought as in the theory of the rainbow, it must once and for all put it on record that by so doing, we introduce simplifications. We instinctively ignore the inessential features and accept certain conceptual distinctions as given, even though they are only approximately realized. This happens especially in the historical and didactic beginnings of a science.

Now we must keep in mind that we do *not, as* is suggested so often, make an inner, conceptual representation by virtue of which certain connections between events become conscious through the fact that they are now formulated *conceptually*. Perhaps this notion is propagated especially by those who, arguing brilliantly, deny any reality to thought other than that it is a representation within the human being.

In contrast, what we do in our thinking about appearances is just this: *We open a second door toward events that take place on the stage of consciousness.* We take hold of connections, we establish regularities, we learn to distinguish the essential from the inessential. In short, we organize phenomena and bring them into coherence with other phenomena. Thus, we gain, in pure thought, the ability *to see more in the appearances than they offer of themselves if we meet them passively.* (Both causality and spatiotemporal measure are pure thoughts.)

After penetrating the rainbow with thought, every detail has its place. Under the *dominance of concepts,* the details develop a profile, become significant, or perhaps become trivial side issues. The representation that has arisen in our consciousness has *reality;* it is not a mere image or reflection, which as such would be just as enigmatic as the raw phenomenon.

I claim nothing less than this: *Reality is not behind things but in them,* although it is veiled to begin with. Thought brings this new kind of reality to *existence* within man. Conditioned by many kinds of bias, this existence often seems merely subjective to people today.

I do not want to say that just any thought relating to phenomena is already real, not even (as it is experienced in various ways with various people) that it must someday be real. We are concerned here with a new approach: *In the things* (more exactly, in the phenomenon along with the descriptive concepts that serve the organizing function of bringing the related things into consciousness), *together with what enters consciousness through the door of thinking, reality can be found.* The reader should have ascertained already that it is concepts and ideas that enter our consciousness through thought.

I still have to refute the objection that our knowledge of reality is only subjective. Before turning to the relation of opposing views, another side of the matter shall be discussed. Thoughts alone are (of course) not reality. Consciousness cannot attain a concrete *sense* content just by weaving concepts, however finely etched they may be. Nor is the world of phenomena a reality when it is merely stared at. The misunderstanding of both the naïve realist and the modern positivist is that both forget thinking. The first forgets it by placing what is added by thought into the world of perception (and actually there are not even *objects* there but only discrete sensory impressions). The other forgets thinking when demanding that the world of facts alone should speak, for the facts of the positivist are clarified and ordered and so on through thinking, but more critically and thoughtfully than for naïve consciousness.*

The statement that thinking cannot produce sense content out of itself does not deny significance to pure thinking. Whoever has not learned to remain *conscious* in the field of pure thoughts (and not, for instance, *dream* into word-sounds) will never be able to become a natural scientist in the sense described here. That is the *true* reason for the great part mathematics plays in natural science and, in particular, in physics. Mathematical thinking is more or less the only field in which, to this day, pure thinking is practiced. The wealth of mathematics contradicts the opinion that pure thinking is an empty ability. It has *content* and can be experienced as a meaningful activity. There can be no doubt that mathematical thinking must be practiced before one can approach the task of unraveling the riddle of natural phenomena with even a small measure of hope for success.

Of course, I do not maintain that every physicist today is conscious of this situation. He learns the methods of pure thought

---

*criticism of positivism should not be taken as a rejection of its great
 ur consciousness of the need to provide a sense-based foundation

more or less mechanically and lives entangled in the dogma of the basic unknowable nature of reality. He does not use the *thinking* that is meant here, yet with that, ever new theoretical concepts of the greatest consequence come into existence instinctively, in a certain sense. For it should be sufficiently clear from what has been stated that neither the methods nor results of twentieth-century physics should be denied wholesale or portrayed as monstrous horrors. What we strive for together here is the *self-understanding* of thinking in physics.

We continue this effort now by regarding the conceptually grasped *natural laws* as separate from the *phenomena*. Of course, these laws too require descriptive concepts (i.e., the ordering concepts with which ordinary things are permeated).

## What, Then, Is a Natural Law?

Before we form opinions of our own on this question, a few characteristic opinions are given here as examples. Louis de Broglie says, in describing the crisis of determinism, that the Chaldean shepherds were the first to find "that the movement of the stars did not take place haphazardly, but obeyed unchangeable prescribed courses; and perhaps they grasped a general thought, perhaps they felt intuitively that nature obeys laws" (p. 223ff).[5]

An account then follows that is not intended as a description of determinism, although it leads to its characterization:

> If one says that there are natural laws it means that the phenomena are connected in an unchangeable order, and that when a system of conditions is fulfilled this or that phenomenon follows of necessity. In the same measure, as man outgrew the age of Chaldean shepherds and better learned to observe the universe that surrounds him, he succeeded in distinguishing in the world of physics a growing number of ever newly confirmed laws. By and by almost all who turned to the study of natural sciences believed, therefore, that the physical world is a huge machine, the course of which is inexorably determined.[5]

*A New View of Nature?*

Following this, he repeats in his essay, "On the Calculation of Probabilities," a well-known quotation by Pierre-Simon Laplace about the all-encompassing spirit:

> If it were possible for someone, in a given moment, to know all the forces that move nature, and also to know the position of the beings of which it consists, and whose spirit would be sufficiently encompassing to analyze the phenomena, he could include the movements of the largest bodies of the universe and that of the lightest atoms in one and the same formula. Nothing would be uncertain for him anymore, and the future, as the past, would be present before his eyes.

De Broglie calls the "sentence rightfully famous by virtue of its precision of thought and elegance of form." Next to the practical value of determinism—that the researcher does not capitulate in the face of the confusion but starts looking for hidden laws—"the deterministic dogma without doubt contains part of the truth. There would be neither order, nor regularity in the phenomena of physics if it were absolutely false, and all science of these phenomena would then be impossible. Physics, however, exists. That is an undeniable fact. It has shown its value by virtue of much progress and numerous useful applications" (p. 224).[5]

And a little later, it is said expressly:

> Even the newest theories (i.e., quantum theories —EDS.), which physicists had to propound almost against their will in order to explain the experimental facts, do not go as far as to eliminate determinism from physics completely.... But they don't see it anymore as strict and universal, they set limits to it. (p. 225)

Then, as a consequence, "the new ideas of quantum mechanics" are described as follows:

> Let's summarize the ideas developed thus far. The reality of physics seems to be formed on a microscopic scale out of beings that show successive incarnations with sudden metamorphoses, incarnations that cannot be described with the

aid of calculus within the framework of continuity and of determinism. The statistical aspect, however, of these kaleidoscopic changes is describable with the artifice of coordinated waves in a classical way.[5]

De Broglie closes this chapter with a question:

> Should one go as far as Bohr and believe that we will understand, thanks to the new conceptions of contemporary physics, why the classical methods of objective science seemingly fit so badly to the phenomena of life and spirit? May we believe that microscopic physics could create mediating transitions between the macroscopic reality of physics, in which mechanism and determinism rule, and other, more subtle fields, in which these conceptions, if not false, yet are not usable?

One can say that De Broglie acknowledges conceptual necessity in the law of nature. He recognizes that changing the interpretation from conceptual necessity in the law of nature to iron necessity leads into determinism, and he sees in the methods of the new physics something that can do justice to the spontaneous and sudden "incarnation of beings." Perhaps, according to a hope of Niels Bohr, it can also do justice to the laws of life and spirit. He does not dare follow the path further but concludes:

> I want to limit myself to my field, physics, and not answer this question. I will only point out that the discovery of quanta, the consequences of which we do not yet see in their full extent, demands a total reversal of scientific thought. This turning point is the most significant that exists in the long space of time, which science had to go through in order to bring into agreement the picture of the world of physics with the demands of our reason, as far as possible.

C. F. von Weizsäcker acknowledges natural law in its kinship to mathematical law as follows. He says in this context that the natural scientists of the Renaissance liked to link themselves with Plato instead of with the torpid and misunderstood Aristotle:

## A New View of Nature?

> For Plato, only pure mathematics has any claim to be recognized as true knowledge, in that the proper claim to knowledge is reserved only for the philosophical knowledge of ideas; about the world of senses one can, even with the aid of mathematics, only tell a probable story. For Galileo, however, mathematical law is rigorously valid in nature and can be discovered by effort of human thinking, to which effort also belongs the carrying out of experiments. Nature is complicated and offers us, not always on its own, the simple situations in which the one law we want to study is working, free of interference by other effects. But these disturbances are caused by forces that obey their own laws; they themselves are also accessible to mathematical studies. Do not tire of dissecting nature and you will become its master. The realism of modern physicists believes neither naïvely in the senses, nor does it treat them contemptuously in spiritual hubris. (p. 110).[34]

Here, contemporary physicists speak *for* natural law. In any case, they speak for the part that remains after the developments of the twentieth century. Let's summarize: Natural law is thought of in accordance with the pattern of *mathematical law* by both its critics, who cannot quite dispense with it, and its defenders. Without question, *necessity* rules for the mathematical law.

A comment about mathematics is needed here. Mathematic, in this connection is *not* merely the examination of logical deductions from arbitrary premises. On the contrary, it is the theory of relationships that, without a trace of arbitrariness, flow with necessity from the most plausibly thinkable fundamental assumptions about certain mutual relationships of the basic elements. We are urged to both by the idealizing worldview of the basic elements and the basic assumptions. It is not being denied that mathematics is a free creation of insight, but it is being denied that mathematics consists merely of drawing conclusions from arbitrary premises. In the character of mathematics I just described lay and lies *its essence as knowledge* par excellence throughout its history, which lasted for more than two and a half thousand years. What has been stated is valid even for those strict formalists who *feel* they are allowed to engage in only an analysis of the structure and not the content of *propositions*. Even they seek a real *experience* of knowledge in their analysis of mathematical structure. For, as Bruno von Freytag-Löringhoff convincingly proves, so-called symbolic logic and the newer "logics" are nothing but "fully intact little pieces" of *mathematics*, which by no means exhaust actual logic—despite the views of some mathematical logicians (p. 155ff).[7]

Of course, a *law* in natural science arising in accordance with a mathematical prototype is not imbued with the necessity that rules in mathematics. If it were, mathematics would be placed in nature and we would be led directly to determinism, as Laplace formulated it. Determinism would be the "fate" of modern thought.

## Natural Law

I will now develop my own views with the hope that the reader will participate. We find ourselves facing phenomena of nature, and in the abundance of regularities, we find *law*.

Everything now depends on whether we can listen to thinking itself while taking hold of the laws. Is it true that the organizing elements at work are relationships internal to human beings, that these are naïvely placed into the world and need to be overcome? The question arises as to whether it is a "real outside world" or a Platonic "world of ideas" in which we objectify the only subjective concepts. Must we be satisfied with stating the fact that we are formulating, with the aid of "laws" alone, certain maxims of procedure? That is, must we be satisfied with no more than "operational rules" that allow *us*, to a certain degree, to *predict* phenomena that follow certain other phenomena, or groups of them, in a time sequence? And is the mathematical *form* only a means to formulate this prophecy *quantitatively*, and thereby more exactly verifiably?

## A New View of Nature?

Erwin Schrödinger is concerned with these questions. He describes the position of the *positivist* regarding the question of *gestalt* in a kind of dialogue:

> And shall we be forbidden to look for it (the gestalt, ed.), and think about it? Should this be as meaningless as a cloud, which looks now like a camel, now like a whale, and yet is neither? "Not at all," the positivist will answer, "on the contrary, that is what you should do eagerly; look especially for the gestalt relationships between the various forms of this kind" (i.e., between the more or less ascertained theories in the various disciplines). That is exactly the way to organize the description more simply, more encompassing, more economical in thinking. But never be tempted to believe that you will ever get beyond the merely external description in this way. You cannot peek behind the scenery. (1261 p. 40ff.)

The positivist would, it is true, admit that we are concerned precisely with a *picture* itself, and that prophecy is merely a means for testing the picture. He would add the proviso that one only wants to understand, without adding fantasy pictures involving unobservables, with which one tries "to explain," as you call it. The positivist then continues (with Schrödinger):

> But I know you; you have a tendency to take and proclaim not the facts themselves, but instead, those auxiliary constructions, as if you had "discovered" them, that is, to take them for the actual gain. And I will not take part in this. For whatever has no direct relationship to a possible perception must be kept out.

Schrödinger then points to the exaggeration of the methodological principle of Mach and the totally faulty prognosis in Mach's rejection of the images of atoms and molecules. "To decry him today for that would be a cheap triumph." But, continues Schrödinger, with a refreshing lack of ambiguity:

> Nowadays, we are told with particular reference to Mach's work that we cannot expect more than these predictions that

are repeated ad nauseam by our science. Down with all "idolatry." Only differential equations, or other mathematical expressions, and a recipe for how one can deduce, from them and a number of actual observations, pronouncements about future observations are all that it is in principle possible to know about in advance. The longing for clear pictures, it is said, signifies wanting to know how nature really is constituted. But that is supposed to be metaphysics—an expression used by today's natural science mainly as a term of abuse.

It is informative to read the inaugural lecture by Schrödinger, republished in the same volume (1922), which is wholly concerned with the battle for and against determinism. Following up on Sigmund Exner, formulations that reflect Schrödinger's struggle for an adequate concept of "law" can be found there:

> The circumstances that precede a certain, often-observed sequence of appearances (A), are divided typically into two groups, *stable* and changing. When, furthermore, it is recognized that the stable ones are always followed by A, then that leads to explaining this group of circumstances as being the *causal* condition of A.
>
> Thus arises, hand in glove with the recognition of *specific* regular interconnections, the picture of the *generally necessary* interconnection of the appearances themselves. A common postulate lying *beyond perceptual experience* is being stated that, even when one has not succeeded in isolating the determining causes of a sequence of appearances, they are still statable in principle. In other words, each natural occurrence is at the very least determined by the totality of the circumstances or physical conditions as they are occurring. This postulate, which also has been called the principle of causality, is confirmed again and anew by the advancing knowledge of specific determining causes.
>
> *Don't we mean by natural law nothing but a regularity that has been established with sufficient certainty in the sequence of appearance, in as much as this regularity is thought of as necessary in the context of the above mentioned postulate.* (p. 10, emphasis added)[26]

## A New View of Nature?

In contrast to this, Schrödinger states:

> Physical science has proven beyond doubt within the last four to five decades that, at least for the overwhelming majority of the sequences of appearance whose regularity and stability have led to the establishment of the postulate of general causality, the common root of the observed strict adherence to the law is—*chance*. (p. 10)

He adds that a single molecular process might not be strictly lawful:

> The observed lawfulness of mass appearance does *not* need to be analyzed in detail. On the contrary, the lawfulness in detail is hidden by the averages of millions of single processes. This is all that is available to us. These averages show their own purely *statistical lawfulness,* which would also be present if the course of each single molecular process were determined by throwing dice, playing roulette, or drawing straws. (pp. 11–12)

By means of an example from the *Laws of Gases,* he continues:

> It would be completely sufficient to assume that, due to a single impact an increase or a decrease of the mechanical energy and of the momentum is *equally probable,* so that on *the average* the quantities *of very many impacts* would in fact remain constant; roughly in such a way as one throws two dice one million times and gets an average value of seven, while the result of the single throw is uncertain.

One can say that the philosophical interest of the physicist was aimed at two major questions during the discussions of the last decades: Is determinism valid in the factual world if one translates the mathematical equations back into common language (and by doing that generalizes them to have absolute and universal validity)? Is the *law of causality* violated if one has to introduce *fundamentally undetermined, only statistically describable,* elemental acts in order to interpret the phenomenally effective procedure of the wave-particle reconciliation?

They were concerned in that way only with the relationships between observation and thought, or equivalently between phenomenon and law, in as far as it involved the interpretation of facts and their interconnections that are difficult to visualize. The researchers touch upon these purely philosophical questions more readily when they are off the record, for instance, in lectures to the general public. This is mostly because the public has been aroused by various popularizations of the "revolution in the worldview of physics."

After evaluating the current literature in detail, one can certainly state: It is true that the concerns regarding the theory of knowledge of modern physics were concerned with methodological factors such as the questions about the justification of idealized objects, the criticism about their objectification, the limits of making them perceptible by aid of analogies from the realm of everyday objects, the new conceptualization of old concepts like corporeal object, or even the question about the consistent identity of "particles," and many others. Yet *one* basic question of the theory of knowledge about the *nature* of thought itself remained untouched.

Everyone is clear about the fact that, to understand one another, more is needed than *words* alone. In addition, the *meaning* of the words—which already has the nature of *thinking*—has to be understood by both parties in a conversation. Furthermore, everyone is clear that those scientific investigations that were advanced with great care had to be carried out with nothing but *thinking*.

However, this (in a sense *initial*) thinking was put up with more as a necessary, because unavoidable, evil. The well-known joke by Niels Bohr, as he was washing glasses in a ski hut, pensively stating that "a philosopher would never believe that one can clean a dirty glass with dirty water and a dirty rag" points to the way concepts crystallize. Crystalline minerals are born according to their nature from cloudy mother water, and *concepts* are likewise formed from unclear ideas according to *their* nature.

Hence, we are directed to the only path not yet taken in the field of natural philosophy, namely, to the one of introspectively examining the content of a concept. This is shown in the above dish washing anecdote to be the inherent nature of the concept.

All of this will be said elsewhere (chapter 2) in another context. I have partly anticipated this point of view, possibly in a provocative way, by saying that "reality" is not *behind* but *in* the things and takes on merely a new form in the thought picture. I now want to continue this thought, and as promised in the beginning of this chapter, elaborate it into a *world picture*.

## Concepts and Laws Have Content through Thinking

To begin with, I will not put this world picture into a *finished* form. Materialism of the nineteenth century strove for a final worldview with the aid of classical physics. In the twentieth century, positivism denies the possibility of such a picture and tacitly denies the objective significance of knowledge, as well. I maintain, however, that the concepts of physics, when they are truly *thought,* are true knowledge in the sense understood here. They remain true knowledge even after their limits are shown later by more encompassing theories, as well as when, beyond those limits, concepts of a new type are needed. In particular, I believe that the concepts of classical physics have already been sufficiently clarified in their mutual relationships that they constitute knowledge. This already suggests that many concepts of quantum physics are knowledge also (i.e., knowledge of reality) if one, as required above, really thinks them.

Now I must show through examples and with the reader's participation what is meant by the expression *"think* concepts." Philosophical language does not provide a better expressions. Here are three simple examples.

> 1. In the elementary theory of planar area, instead of area we could speak only of the number of unit squares that belong to a figure. Having established a square grid, limits could be used to obtain areas of forms that do not consist of an integral number

of unit squares. The limits would arise from progressively refining the density of the square grid. Thus, it is possible to assign a *numerical measure* to each form and gain, by way of abstraction, the concept of area. The procedure is mathematically possible. For certain purposes, it is even unavoidable. I will not deny that the actual concept of area is based in perception to a greater degree than the theory suggests. The relationships of measure theory are already known *before* the notion is rendered secure by the introduction of limits. Naturally, geometry must not be satisfied with the *intuitive* quantity concept of an area. An area is *really thought* about; yet it is also *experienced,* to use an unfashionable term, as an intuitive quantity concept (i.e., thought of as containing substance). To do without the intuitive notion would mean to lose its true geometric meaning.

2. The concept of the mass of a body *can* be defined as the sum of the inertia of fictive particles—*really thought*. In elementary mechanics, it belongs to the *basic concepts* of *measure*. This would be true even if one were to understand mass merely as a coefficient in the law of force acceleration. (compare p. 28ff)[31]

3. To establish the concept of heat, in addition to *intensity,* which is measured by temperature, there is a need for *extensity*. This is provided in the usual thermodynamics through the concept of *heat* as a form of energy. Entropy—an indispensable *quantity* for the examination of the processes in the non-equilibrium case—is added to these. In phenomenological thermodynamics, entropy is frequently introduced as an integrating factor of the differential of *heat quantity* (divided by the temperature at which it is transferred)—a very elegant process, speaking mathematically, with the disadvantage that all perceptibility is lacking in the quantity thus defined.

It becomes more "perceptible" in kinetic theory for measuring a definite probability, although the basic hypothetical totalities are unobservable. Entropy can be interpreted as proper thermodynamic extensity. However, then it is a new kind of "heat quantity" and is thus given an intuitive *content* (compare p. 110ff).[31]

Also, it is a gain for kinetic theory to have beside the statistical interpretation another interpretation for entropy that has content.

## Cognitive Character in Concept Extension

To form a picture of the world as it appears in physics, we followed together a certain guideline. At the first step in a scientific observation of nature, which consists of the description and formulation of phenomena, there already takes place a complete cognitive penetration. We do not observe a disconnected mosaic of sense impressions, but we direct our attention to forms and connections of all kinds and bring concepts to those by thinking. We organize the phenomena, too, according to "regularities" that were stressed by the different authors quoted here, and we attempt to raise empirical rules to laws of nature. We form these laws according to mathematical method. In other words, the primary form of the law, which must be worked out scientifically in thought, is to provide necessary and sufficient conditions for the occurrence of phenomena. This is the second step of scientific observation of nature.

Given the entangled nature of initial data, we might not expect the laws gained from it (also in combination with knowledge gained in the course of scientific development) to be valid in an absolute sense. On the one hand, the observations are of limited precision; on the other, they are of limited significance, for we don't initially know if the observed characteristics are essential (a typically cognitive determination).

In many cases, the given data are already concepts with contents that have arrived in our consciousness through that "second door." They also call frequently enough on understanding for the formation of further concepts. If we consider—for the moment—those concepts already in existence as represented by picture images, the concept that we seek adds nothing to those out of itself and yet changes our whole attitude with regard to them.

Expanding the relationships between concepts, by such a newly "conceived" or "formed" concept, seems to be the aim of the scientist who may have wrestled for years with an enigma. I will provide here two examples from the history of mathematics in order to show the degree to which a "correct" idea depends on viewing relationships in an entirely new way with newly experienced content. The so-called problem of parallel lines is the first example; transfinite numbers are the second.

## Parallel Postulate as an Example of the Way an Axiom Is Experienced ("Actually Thought")

Mathematicians have attempted to repair the assumed defect in the structure of Euclid ever since he formulated the parallel axiom (depending on the version—the fifth or eleventh postulate) in such a way that it seemed an embarrassment when used in a proof of a particular theorem. What is the issue from our present point of view?

First, we will recapitulate. Euclid used the postulate relatively late in his treatment and thus let it appear as a proposition that might be derivable. "In itself a true theorem, but unjustifiably, so to speak, made into a postulate from lack of proof," so the commentators judged. To be sure, the character of the axiom or postulate was under discussion from the start, but the real solution could not be found until much later. Euclid was correct in placing it among the postulates. Because meaningful axioms are unprovable, it must be possible that the opposite (negation) of an axiom *is possible*. Otherwise, by virtue of the inconceivability of its opposite, the supposed axiom would be proven.

Thus, there were some geometers who, while attempting to prove the statement in question, took the route of indirect proof. Hoping to find an inner contradiction, they ventured into the broad expanses of the new territory that arose out of inferences from denying the parallel postulate. They did not yet have the concept that they were in a world of *true geometry* that, however, rested on an unfamiliar postulate. Yet these geometers established many coherent concepts so that, in the end, a solution was found that appears in retrospect like the egg of Columbus: The proposition of Euclid is not provable at all because it is an axiom. Note well: The fact that axioms are not provable belonged to this definition from the start. It is one thing to say this, another to experience it with inner conviction. Therefore, the finding of this solution was by no means the result of a lucky intuition. As in almost no other example do we see here that the real issue is not the concept: *axiom = unprovable statement.*

This shows the path only to the content of the concept. Here it matters most that this latter is experienced concretely in a new context. For one needs only to ask: Until the modern age, what was the obstacle to taking the parallel postulate seriously as an axiom? One must say it is the fact that Euclidean geometry exists and that its truths were experienced as intuitively evident.

The problem within the framework of this existing geometry seemed merely to be how to bring Euclid's postulate into relationship with the other evident postulates. The consequence, however, of the non-provability of the parallel postulate is that, besides the known geometry, there must exist another, internally consistent geometry, even if it contradicts the way of perceiving trained by Euclidean geometry.

It may be made clearer with the help of another example, from the area of pure thinking, what it means to ponder certain meaningful thoughts anew with all one's power of spirit and soul. The new theories in geometry are meaningful and true, even if the earlier perceptions, which specify the sense experiences represented by geometry, have to be given up. Perhaps it becomes clear just with this example that content-filled (i.e., meaningful) does not necessarily mean "intuitive."

## The Creation of a Concept (Transfinite Numbers)

Since Bernard Bolzano, the difficulties in dealing with the infinite were known under the heading "Paradoxes of the Infinite." They did their mischief for a while through faulty proofs, as well as in *sophistic* contradictions. An important consequence in the development of calculus was reached: One may not operate with "infinite quantities" as is done with the finite ones. The extension of the natural numbers to whole numbers, negative numbers, rational numbers, and ultimately imaginary or complex numbers succeeded, last but not least, as a consequence of the power of a symbolism that included all of them. This power of symbolism, so to speak, functioned under all these extensions. However, the arithmetical symbolism failed in the face of "infinite quantities." They proved to be non-*concepts*. Yet, in the hand of Georg Cantor, there was a certain thought that was a key for the creation of "actual infinite" powers (as he called the numbers) differing from one another. This multiplicity of infinites was a thought that would have been even more paradoxical for anyone else and can therefore be taken as an indication of the outer limit for the thought of that time.

The essential element for the creation of the transfinite numbers lies in the following steps of thought. A concept of equality that also embraces infinite sets, called "equivalency," must be explained, and one must show, especially among the infinite sets, that there are sets that are not equivalent. The equivalence extends the concept of quantity (cardinal number) to that of "cardinality"; the existence of non-equivalent sets enriches the world of orders of magnitude—by way of the transfinite numbers.

On the one hand, the key thought is the renunciation of an axiom in mathematics, and on the other hand, drawing conclusions from the generalized concept of cardinality. The principle "the whole is larger than the part" cannot be maintained for cardinality. A set can therefore be equivalent (of equal power) to one of its partial sets. This, however, is possible only with infinite totalities. Thus, for instance, the set of natural numbers and the set of even numbers are equivalent by way of the correspondence $n \leftrightarrow 2n$. Were one to say, naïvely, that because of this correspondence there existed equal quantities of natural and even numbers, a critic could argue that the natural numbers included also odd numbers so that there must be more natural than even numbers.

Georg Cantor bases the concept of equivalence only on the possibility of creating a correspondence. Therefore, categories such as "equal" or "larger than" must not be applied in the usual way to this concept. Despite this, there is still justification in speaking of non-equivalent sets. Those are the ones where *no* correspondence can be created among its elements. The progress that Cantor brought about lies in showing that a 1:1 correspondence between the elements of certain infinite sets is not possible; namely, they are not equivalent. Since it is clear in the simplest case (the set of integers) that one of any two such sets is equivalent to a subset of the other but not the other way round, meaning is given to an extended kind of comparison of quantity of infinite sets. That is the way actual infinites are created.

The concrete execution of the proof that two specific infinite sets are not equivalent can therefore only take place when it can be shown that every thinkable one-to-one correspondence is led ad absurdum.

This is not the place to follow up on the backlash or its consequences, which Cantor's theory of sets suffered at the beginning of this century. The affair was a shock to confidence in mathematical logic. It may be sufficient to point to a strange parallel: From other, quite different sources, confidence in the power of human thought was lost during the first quarter of the twentieth century. For physicists, this loss was caused by the impossibility of perceptible models in the face of abundant guaranteed new facts. For mathematicians, faced with seeming contradictions as a result of overly audacious attempts at forming concepts in the theorem of sets, one doubted the adequacy of mathematical logic.

## Does a Picture of the World Exist that Arises from Thinking that Grasps Reality?

We now address an objection that easily could arise against our attempt to design a picture of the world by means of *reality encompassing thinking*. For if thinking makes the claim to conceptually encompass reality as a complement to sensory experience, is not then a certain physical theory, provided it is correct, already the reality?

One can reply that it is only a prejudice arising from the previously overcome objectification of a *real world,* in the sense of naïve realists, that causes us to expect that this reality is self-contained and unchangeable and that what changes is only our respective degree of insight.

Whoever has thought through the critique of simultaneity by way of the theory of relativity will not be offended by the criticism of the concept of reality just expressed. To the degree that any kind of simultaneity in the theory of relativity is being denied, are we losing, through our restriction, everyday "objective reality" if we maintain that objective reality has no meaning "unless we think it."

Thus, a certain theory encompasses a part of reality. Another, higher theory that encompasses many elementary laws from a broader point of view must not, by that, degrade the elementary laws or theories to mere approximations. Let's look at this using the example of optics. The theory, in its first stage, postulates the rectilinear spreading of light effects, that is, rays (to be more exact, the light-shadow border). In being content, in a justified positivistic spirit, straight lines as an idealization of the *possible* light-shadow borders expresses one side of reality. A ray of light is straight in the same sense as a straight line drawn with a ruler. My intention in drawing corresponds to the natural law governing the ray at this stage of observation. With that, of course, no anthropomorphism is implied and imposed onto nature.

The point of comparison is only meant in this way: The intention to draw a straight line is of an ideal nature, and so is the law (original phenomenon) of rectilinear light propagation. A perfected theory brings the natural colors along the border seams of light and shadow into connection with observable deviations of rectilinearity through bending. This theory encounters in its mathematical description certain phenomenologically determined lengths that are then ascribed to the colors. Thus, an elementary theory of waves of light comes into existence. It describes the totality of behavior of light in a more encompassing way than does ray optics. At the next level, the transversality in the optics of waves must be taken into account. At yet further levels (Clerk-Maxwell equations, photon theory, quantum conditions), the theory becomes more complete and ever more suitable to gather the widely different areas under a unified perspective.

But neither the straight lines at the first level nor the wave movement at the second level is *the* reality. We need not even conceive of a *reality* patterned after the sense-perceptible objects representing concepts on a third or fourth level. Each of those features that are recognized at their respective stages is part of the world that, in phenomena and thought, is being reunited in consciousness. Such single features of reality contradict each other nor more than the light-shadow border (taken from experience and idealized) contradicts the notion of a straight line, in higher theory, as a perpendicular to wave fronts.

Proceeding from here, we want to grasp the reality of single physical entities within the concepts belonging to them. I will begin with a temporary argument about the concept of matter and then turn again to the method of knowing.

## Matter

By going in quest of *matter as phenomenon*, we detach ourselves from the remnants of a misunderstood concept of matter stemming from the mechanical view of nature. What extends spatially may be called matter, when such attributes as weighability, impenetrability, chemical, electrical, thermal, and still many other qualities belong to it. Consequently, we state as a conclusion of various prescientific experiences that if an object has *some* of the enumerated *material* qualities, then *all* these qualities exist in it. This must be qualified a little. Gases show no impenetrability; inert gases show only traces of chemical qualities. It is a matter of empiricism and of practical natural science to see the lack of one or the other quality in a lawful connection with the other qualities. If, however, phenomena arise that, fundamentally, no longer show the features of matter, understood in that way, then they belong phenomenologically to the boundary of the material world or in a realm lying beyond this boundary.

Schrödinger discusses just such phenomena and arrives at brilliant formulations (p. 121ff).[26] There is only one consequence he does not express clearly: Physics with its modern results moves beyond the borders not only of the material world but also of the sensory world altogether in many different places.

In anticipation of the following chapter, we may contribute something to a total picture.

## Quantum Physics Requires a New Concept

We live in a world of sense impressions and sensory perceptions. Recent research shows that our sense organs are constructed in such a way that their sensitivity almost reaches up to those frontiers that are characterized by the quantum phenomena. Large parts of the sensory world appear to us in such a way that we can recognize global patterns that are independent of the texture of the substratum. Such patterns are meaningfully connected. (We have developed from them the elementary phenomenological theories of physics.)

In biology, we are also led to meaningful connections among our observations. These relationships appearing in living matter are evidently of a different nature from those of the lifeless. Walter

## A New View of Nature?

Heitler, speaking as a physicist, has pointed with special emphasis to this difference.[12]

In the past, there were frequent attempts to derive the patterns of the sensory world described here as macrophysical phenomena from what are assumed to be the real conditions in microphysics. In keeping with to our discussion, the microphysical conditions exist at or beyond the frontier of the material world outlined in the previous chapter.

The phenomenon matter is a thing; we have determined the concept for it. The boundary phenomenon, namely, particles as fragments at the frontier of the material world, is another subject. That leads us to the question: What concepts are required in addition to known mathematical descriptions of quantum phenomena? Expressed differently: How do concepts have to be constructed or, if necessary, recreated?

The example of Cantor can guide us: The apparent impossibility of a new concept of infinity was the key for something new, namely, the transfinite powers. Thus, the impossibility to encompass wave and particle in one concept must become the key for new kinds of entities. Louis de Broglie speaks inquiringly of "beings that are capable of sudden incarnation."

## NEW PHYSICS REQUIRES MATHEMATICALLY IDEAL ELEMENTS

To recapitulate briefly, consider the difficulty emerging with the new phenomena. A *particle*, initially imagined to be physical, shows a definite track and seems to have speed. However, confronted with a double slit, the particle behaves like a wave. It cannot possibly run through both slits as a *particle*. We take it for granted that a wave could do so. Next, interference patterns appear behind the double slit when many particles are intercepted. At the beginning stage of the new physics, the well-known paradox requires a particle with wave properties or a wave that behaves like a particle, depending on the form of the initial question.

Taking our description of the material world as an aid, we state: Therefore the particle is *not an object of the material world*. This requires the creation of a new concept.

Let's take another look at the development of Cantor's ideas. One had to forego thinking of infinite powers as *quantities*. It is a prejudice to hold fast at all costs onto the axiom of quantity (the whole is larger than the part). If one wants to have a concept of infinite powers, one must forego this aspect of the mental picture of quantity. Therefore, it is possible to form a new concept for powers that is not subject to contradiction because properties of quantities are not required of it. The paradoxes of the infinite disappear as a result of the insight that they are only an expression of attempts to think in inappropriate categories.

Using other examples, one can strengthen one's confidence in this means of forming concepts. Let's observe the introduction of infinitely distant points into projective geometry. One has to *change* the content of the concept of the intersection of straight lines if one holds on to the thought that a bundle of straight lines meet at a point, even if the lines are parallel. It is clear that one must *not* assign a point of intersection to the Euclidean parallels of elementary geometry. The parallels remain a fixed distance apart and do not come together arbitrarily closely, which, however, would be necessary for intersecting.*

One can see that in the kind of thinking used in projective geometry, something similar is being executed as in the case of the introduction of transfinite powers. With the latter, one foregoes the character of quantity; with the straight lines running together in ideal points, one foregoes the measuring of distances or angles. In short, the new concept does not possess all the attributes of its forerunners—the mental pictures from which it is formed.

By returning to the particle discussed above, it is only a negation to say that it is not an object of the material world. What, then,

---

\* This is an obvious argument. Axiomatically speaking, it must be expressed somewhat differently.

is a *particle* in the sense of newer physics? In short, and perhaps open to misunderstanding, *it is a mode of the sensory phenomenon of something not physical.*

By that we mean that something at the boundary of the material world can have some material properties without having to possess all of them. The thing in question is fully described for physics by the psi function belonging to it. This is the "Copenhagen interpretation." At a certain stage of the theory, we can follow it.

It may be stressed that one must not a priori exclude the possibility that later experiences may force one to modify the present form of the theory. Such an alteration will leave the main features unchanged, as must be the case with the evolution of a well-established theory. Then the features may remain valid as an approximation to a refined analysis. We have already discussed the way theories capture part of reality.

By taking up such an expanded concept, the pros and cons of the statistical interpretation are not laden with disputes about the "lawfulness of nature" anymore. Overcoming determinism meant and still means that we struggle with the problem of the lawfulness of the world. But Heisenberg's critique in the first third of this century showed that the deterministic view was nothing more than the transference of an idealization, justified in its own sphere, into an imaginary pseudo-reality. This does not make classical physics false by any means. Furthermore, I am not concerned here with interpreting classical physics as a mere approximation of another, more complicated reality but rather with acknowledging it as the true expression of definite sense relationships in a field of phenomena. Therefore, what in another context is often called macrophysics or the macroscopic point of view may, within its own framework, be taken as *knowledge* appropriate to obvious material appearances with few limitations.

This sensory world, however, points beyond itself. The appearances themselves lead the researchers to the boundary previously described.

We have reached a temporary conclusion of this discussion: A picture of nature that takes the classical as well as the newer physics seriously. Using examples, we tried to show how we may hope to use the method of extending the concept, which we found in nineteenth-century mathematics. This outlines our future path: The discussion about the potential extent of human thinking has to be deepened; the arrival of the new physics must be observed in more detail; finally, the separate, concrete formations of concepts will have to be examined more exactly. These tasks will be taken up in the following chapters.

## Chapter 2

# The Method of Gaining Knowledge

### The Role of Thinking Revisited

We must gain a more intimate view of the role of thinking. We need to have some knowledge of thinking to do so, because the notion of *matter* is chameleon-like. Moreover, thought might be used even to deny the possibility that thinking itself can take hold of the essence of things.

Let's try to take thinking itself *as an object* of observation before entering into any epistemology that, after all, employs thinking. We would then be in a position to evaluate the epistemological theory, its insights into the nature of cognition, and also, its significance for the results of contemporary physics.

### Cognition and Thinking

The world meets us as pictures from percepts and out of imagination. This statement is not to be understood in Arthur Schopenhauer's sense, namely, that the world is our imagination. It is a question, instead, of considering that elementary thought activity that consists of taking hold of the names and concepts that present themselves to us. Contemporary physiology of the senses has come to the insight that the *totality* of things seen is given to us in our consciousness and not merely in elementary, immediate sense impressions. Only after careful reflection do we learn that the *primary* impressions are those of colors in various degrees of intensity and saturation. Even here, the question emerges as to whether or not this two-dimensionality is a conceptual conclusion arising from the physical nature of the eye. Nevertheless, the direct percept is always three-dimensional and thing-like, independent of any way that one can "explain" this fact.

Similar things could be stated for the other senses. Yet such a physiology and psychology of the senses cannot be undertaken before an examination of the role of thinking, because these disciplines themselves make use of and thereby presuppose it.

Equally, thinking must not be viewed as a function of matter a priori or, even worse, as if this gave an explanation for its potential to encompass matter. Even if this thesis of the material nature of thinking could be true, it could not be justified *in that way*.

If one walks around the periphery of the problem of *thinking* in this way, one is necessarily led in this preliminary phase to make thinking itself into the object of examination, just as I tentatively suggested. One could call this opening phase a *meta-epistemology*.

Next, it is appropriate—still meta-epistemologically—to become clear about the guidelines of the examination. These are not *binding* in as much as they may not predetermine the result of the examination. Should new perspectives that demand a revision of our methods arise during the progress of the examination, there is no reason to reject them on account of the initial guidelines. Equally, the *natural-scientific method,* to be treated later, which I take now to be our method, must not be misconstrued by saying that epistemology is a natural science.

## Natural-Scientific Method

The natural-scientific method can be characterized in this way:
1. A result of sensory perception is valid, independently of any ideas or theories.
2. Rules and the laws governing events are "idealized."
3. An experiment is a repeatable *observation,* made under freely chosen conditions, and becomes *experience* when it is saturated by the thoughts belonging to it.

A few examples will serve to illustrate the applications of these characteristics. Concerning 1: When we have a measurement in practical natural science, it may be *corrected* only on the basis of the same facts and on the basis of possible connections that have already been worked through in thought. There are innumerable examples of such justified *corrections*—correction of a gauge pressure to get atmospheric pressure; correction of the reading of a thermometer for the fact that the liquid in the stem is not immersed and is therefore not at the measured temperature; correction of the astronomical determination of the position of a star on account of atmospheric refraction. However, an observation may never be ignored because of *impossibility* or teleological arguments.

Concerning 2: In most cases, *idealization* means that, apart from the "accidental" disturbances, one brings "simple," mostly mathematical relationships to bear upon the observations. These are most often in the form of *if-then* pronouncements.

Concerning 3: Through experimentation in particular, observation rises to the level of experience. In principle, the experiment must be repeatable. The essence of the art of experimentation is the use of all pertinent measures emerging from the currently achieved level of empirical and theoretical knowledge. These all have to be taken into account to make the answer conclusive; otherwise, we gain an "artifact"—that is, an observation conditioned by aspects of experimental conditions that are not fully known instead of a result experimentally reached. Such observations are justified as simply that—*observations*—but they may be scientifically invalid through erroneous interpretation.

Here is a brief (and thus incomplete) summary of the principles of experimental natural science, consisting of

1. the primacy of experience,
2. the rational order of experiences (leading to *theory*),
3. and the experiment as a source of experience and as a corrector of theory.

Experience and logic are the final arbiters. Experimental results are unreliable so long as they are not repeated. A theory is unsatisfactory so long as it shows inner contradictions.

Now I return to our initial point of departure, asking: What are the results when the principles of natural science are applied to *their own* foundations?

## The Experience of Thinking

Thinking is at hand for immediate experience as the object of observation, like any other content of consciousness. This is often overlooked, or its potential is contested, because the actual thought process is not easily observed. One knows, however, the results: ideas and concepts. After they have already emerged, the observation of ideas and concepts teaches those who want to open themselves up to it at all something that involves both observation and thought together. When observing thought—and only then—the observation and the activity of thinking are of the same quality, even though not necessarily occurring simultaneously. One can experience something that cannot be denied, just because it is *experience*. It has to be respected according to the natural-scientific method. It is true that this experience that I am going to discuss is of a kind that everyone can only have for themselves, but it proves to be the same for everyone, as will be described below. There is a contrast here with the experiences of scientific experimenters. *They* have a certain experience that we believe in, because, in principle, we could have it ourselves. With the experience described here, concerning thinking, the fact is that we *must* have it ourselves. Its importance lies in this fact, because they have the experience themselves; they are also sure that nothing can creep in to adulterate it, creep between them as the experimenters and their object—*their own thinking.*

Everyone *can* have this experience, and *must* have it if one wishes to be competent in such matters. It consists of this: I observe the *content* of a concept that appears as belonging to a thing. This

## The Method of Gaining Knowledge

concept does not pose a question; rather, it answers the *what* of the thing observed. It is clear that this answer does not need to be exhaustive. On the next level, facts and concepts gained in this way are again the *things* for new answers. In short: The content of the concept is not in need of explanation at its emergence. It is rather, itself, illuminating other things.

One can have this experience. Rudolf Steiner pointed to it.[28] No one had spoken *in this way* about the content of the concept prior to him. Many have confirmed the correctness of this central observation through their own experience.

Anyone can do it who has sufficiently good will. No one can take away this certainty of experience, which is also knowledge. Every doubt of Descartes—whose questions led him to say *cogito, ergo sum*—is silenced at this point. Nonetheless, the certainty to be won here does not guarantee eternal duration of one's own self any more than *cogito, ergo sum* does, but then that is not the aim of this self-observation. The aim is to reach the certainty that the content of a concept is determined by the respective object, but it does not depend on one's arbitrary action. The content is something that transcends one's subjectivity.

In the central experience of thought as just described, the question—*Is knowledge possible?*—answers itself with a strong *yes*. Within the fact that something that transcends appears within me, the seemingly unbridgeable contrast of subject and object is alleviated. This contrast, established by thinking, cannot possibly be any deeper than thinking. It is determined by thinking and therefore is a concept. It exists rightfully from a certain point of view. The concept of this contrast merely explains the other observation: There are inner experiences that depend on me (the subject) and some that are independent of me and adhere to the object in the outer world. The contrast indicated *cannot* emerge when the knowing subject and object are as close to one another as possible—i.e., as it is with the central experience. The transcending nature follows from this experience. We now call it the *universal nature of thinking*.

Let me summarize. The observation of thinking can teach us that the content of the concept of a thing belongs as much to the thing as it does to the thinker. Through the *realization (perception) of the idea in reality,* that reality is grasped by the one gaining the knowledge. Therefore, knowledge of the outer world is possible specifically because this world appears not only to perception but also to thinking. This is solely a result of introspection. It is the only result that has an assured character of experience, and it must become the basis for everything else. Therefore, it is important to follow up with examples of the fact cognition means to find a concept and to show how this fact appears in practice in scientific thought.

When one chooses an example from the realm of pure thinking, one surely does experience the certainty of cognition, but nothing is gained yet with regard to the *universality* of thinking in relation to the outside world. Further examples must therefore follow.

Frequently, an insight into the general validity of reasoning is achieved in mathematics. This occurs in such a way that one realizes that, without limitation of generality, a certain element may be chosen in a particular way.* It is completely clear in each case what is meant, and it only occasionally needs a minor clarification. The particular means of expression used has no bearing on the mathematical proof. It serves merely to draw the attention to the fact that the concrete case has *general* validity. I know of no better example in the sphere of pure thinking than to observe in oneself or others how suddenly the concept of general validity lights up. It would be a complete misunderstanding to deny this experience by saying that one just *generalizes.* Premature generalization is definitely the mark of the untrained intellect. It uses the *form* of the transition from the particular to the general, without accompanying this with the experience, which, after all, brings the content of a concept (general validity, for example) to consciousness.

---

\* For example, a line through the center of a circle intersects the circle at two points. Without limitation of generality, the line can be taken as vertical.

What are some examples from our experience that the content of a concept is independent of the subject who thinks it? The concept of heat, which is not yet fully clear, has to do with the state of a body being warm and also the quantity of heat. It was only in part fixed by the invention of the thermometer. The observation that the thermometer upon immersion into hot water sinks slightly to begin with, and only then rises, can exist as a single observation; then it is a curiosity or a paradox. Yet, if thinking connects the generally valid actual experience of spreading heat with another general idea, namely, that of expansion by heat, then this thought can arise: The heat entering the thermometer expands the container first and then the contents. With that, the initial lowering of the thermometer at the moment of immersion is explained.

The seeming contraction arises from the increasing volume of the container. Many thoughts rally together, in a certain sense, with such an explanation. As iron particles orient themselves under the effect of a magnet, likewise the previously disconnected single thoughts, which had been gained earlier in a more or less clearly formulated way, organize themselves—spreading of heat, expanding effect due to heat, and heat propagation requiring time. The use of causal interlocking joins these.

To stay with the thermometer...the fact that it actually shows only its own temperature is so basic that we don't think about it. This can, however, become a step in cognition if one wants to take hold of the concept *"temperature"* more exactly. This rests on the experience that bodies in contact equalize their warmth condition, a condition that one can pursue in principle with any properties of varying warmth. The thermometer stands in the place of the properties. The basic transitivity of the equilibrium is tacitly presumed.

Let's continue by pursuing the formation of a concept as important as that of electric fields. In the mid-nineteenth century, many separate electrical phenomena were known. And in the seventeenth century, the *electrification* of bodies by friction led William Gilbert to the expression *vis ekctrica* (Greek *elektron* = amber).

Otto von Guericke discovered *induction*. The eighteenth century brought the discovery of resinous (and vitreous) electricity and the difference between conductors and nonconductors, as well as the principle that opposite kinds of electricity always manifest in equal quantity. As helpful auxiliary images, a dualistic and unitarian theory was developed in the same century. Charles-Augustin de Coulomb influenced his contemporaries at the end of the eighteenth century mainly through his experiments with the torsion balance and the half-intuitive postulate of the law of attraction that bears his name and is analogous to Newton's law of gravity.

One could have started from a theory of the potential, as it was developed by Green in the nineteenth century, analogous to the potential theory of gravity by Siméon Denis Poisson, and arrive at a mathematical description of the field. Instead, the very plastic field pictures of Michael Faraday had their effect. For him, the field was not an abstract expression of surfaces having the same potential in space; rather, it was what actually moved the charged bodies or exercised forces upon them. The picture of the field lines that tended to shorten themselves was intended as a picture that could be grasped sensorially. It helped to bring the picture of the field to general acknowledgment.

Through Faraday, the concept of *action at a distance* was overcome (through a theory of continuous action). Using the language of modern formulas, the quantitative connection was established between this imaginative conception and the mathematical theory of energy localized in space. With that, however, the quasi-sense perceptual character of the field receded somewhat.

What *thoughts* has one attained? What value do these realizations, gained in this way, have? One has learned to bring the auxiliary, pictorial idea of the number of field lines into the form of differential equations and to prove as a mathematical theorem that, for a given boundary condition, there is a unique solution describing the field. If one also takes into account the qualities of the so-called dielectric (not gone into here), one arrives at a full conceptual mastery of all possible situations, solely in terms of fields, including that of the equilibrium of charges on conductors. Having done this, all that is essential is seen in the *field;* it is not essential to think of charges as the actual *carriers* of electricity.

In any case, it is possible to sum up the whole theory through superposition of real and hypothetical charges onto a field. One sees that there is cognitive value in the quantitative mastery of the conditions. The *thoughts* that one forms are more effective in the form of relationships than in their connection with the sense-like content of pictures. Herein lies the great significance of Faraday's picture of fields that led to the mathematical relations. It has become a truly adaptable instrument for the rational description of nature.

We can now look back at the character of these thoughts. There might be some arbitrary choices—one might place more stress on fields or more on charges, but the relationships are so encompassing that they include the totality of charge and induction phenomena, Coulomb's law, and the discoveries of Faraday. We experience in this the transcending nature of our thinking. At *this stage of the theory,* we can therefore speak of cognition that reaches into reality. The sum of isolated sensory experiences is therefore no more reality than the empty mathematical structure that is *only of an ideal nature.* We know well that, according to relativistic electrodynamics, the theory just described is an approximation; the form of the fields definitely depends on the motion of the coordinate system. That exactly is the difference between our *realistic view of thinking* and the common nominalistic view that sees in any theory a *practical* formalism and no more. This presentation is in accordance with Rudolf Steiner's discussion of those elements of physics that are not immediately perceivable. He says:

> In the field of experimental physics, it is sometimes necessary to speak not of elements that are immediately perceptible, but of unobservable quantities such as lines of electric or magnetic force. This can also distract us from the unprejudiced

observation of the relationship described here between the percept and the concept worked out in thinking. (p. 123)[28]

The role of sense percept (i.e., the object of perception) is played by the fields. The role of the conceptual side of cognition is played by the mathematical laws. Knowledge results.

Yet something more can be read from this example; it has significance for our upcoming discussions. As electrical and magnetic phenomena are not manifest in the same way as form, warmth, and light are, the representative pictures are *suggestive* but not saturated with sensible content. If one is clear about that, students may be led to a saturated picture of the electrical and the magnetic field as a *space filled with force,* even in the first lesson about electromagnetism.

Looking again at the way thoughts arrange themselves, we notice that it depends on us *whether* we understand the phenomena in question or even whether it becomes a problem for us. As our questions change, the thought connections change, too, but the central connection in thought—what explains, for instance, the paradox of the initial falling of a thermometer, as discussed earlier—does not depend on our arbitrary state of mind. Thus, the naïve thinker, especially in natural science, is not conscious of the conceptual character of his statements but believes he only pronounces on the properties of things.*

---

\* The thinker trained by Immanuel Kant's *Critique of Pure Reason* believes that thoughts take place only in one's head but immediately ignores this stance when applying the results to the world.

# Chapter 3

# The Transition to Twentieth-Century Physics

## What Actually Happened

The images from theoretical atomism of the nineteenth century proved increasingly inadequate in the face of new phenomena. Step by step, changes became necessary. Interestingly, there were things and phenomena, like energy and light, whose continuous nature had not previously been questioned. Now they had to be thought of atomistically. The very way in which discrete quantities were conceived (quantized) necessitated that the older concept of atoms had to be given up. Max Planck, with foresight, avoided the term *atom* and instead substituted *quantum*. Hand in glove with this was the fact that discontinuous *phenomena* that were sought (as in the case of Brownian motion) and partially found (as in the case of alpha rays) were of a kind that did not truly confirm the atomistic picture of the nineteenth century. Rather, they presented increasingly greater difficulties for the imagination. They led to changes in the classical picture of the atom, which were made reluctantly and hesitantly.

One can ask oneself what a picture of nature would have been if scientists had immediately drawn the conclusion at the turn of the century: *Atomism of the nineteenth century is wrong*—not, as happened, *insufficient*. I will come back to that question.

To begin with, the idea of the older picture could still be retained—namely, that ponderable matter is composed of individual *building blocks.* Their way of aggregation would cause the known forms of phenomena of solid, liquid, and gaseous states of *matter.* These particles are no longer *indivisible.* Apart from that, however, one did not doubt their individual existence and fundamental characteristic of being localizable. The new factor was mainly certain kinds of *instability* whose existence is brought to mind by radioactive decay. In addition, the various energy states of the atom revealed in the mutual interactions with light and other kinds of rays were new discoveries. Many expressions that remind one of the older images were subject to incisive changes of meaning.

One learned to describe the states and probabilities of atomic transitions by mathematical methods. These methods can also be used, however, where something other than states of atoms are the concern. For the newly discovered *particles,* called *elementary particles* in a continuation of the earlier usage, could transform themselves into each other according to rules applied to energy states. Thus, they themselves are *states,* and no one can identify states of *what.*

Finally, theory has more recently been preoccupied with postulating systems of equations in such a way that characteristic numerical values for all elementary particles, including nuclei, appear as eigenvalues. The ideal is to let only a very few experimental values—the so-called universal constants—enter fundamentally into these equations.

So far, so good, but an objection can be made: The building blocks have merely lost their indivisible character. Otherwise, they remain, now as before, building blocks out of which all matter is constructed.

The idea of construction, however, has to be corrected as well, for we are not concerned with individual objects, being side by side—so to speak unchanged—in the pattern of things of everyday experience. Certain "features" of the everyday thing-ness are retained; it is true. Yet, others contradict the superposition that is the essential mark of physical *side-by-side-ness.* Think, for example, of the binding forces of homopolar (ionic) bonding. These forces do not belong to the single things in the same sense as, for instance, the forces of cohesion belong to solid bodies or to the fields of magnets. Through the coming together of the particles, a system arises that has altogether new qualities. Additional qualities of solid or liquid bodies also arise.* The situations at lower temperatures already give rise to theoretical pictures that do not let themselves be categorized in a concept of an aggregate constructed of particles. In addition, other phenomena occur, which need a quantum physical description. The behavior of super-cooled helium is an example. The theory of Lev Landau, therefore, can grasp the conditions only when we think of the "passage" of helium atoms through the finest of capillaries not in a real way but in a symbolic or virtual one.

Finally, there were certain *degenerate states* of crystals at low temperatures, which led to the idea that the totality of atoms of which the crystal consists, from a certain point of view, forms *one* quantum-mechanical system. This takes place insofar as the exclusion principle is seen not only for the sum of the states of an elemental particle but also for the variety of states of the whole macroscopically extended construction. An important new question arises from this: How do the single members of a system "know" of each other and the states of the others to avoid occupying the same state several times over?

I have touched on these individual questions in this concluding observation only to formulate the thesis to which one is led by following and finalizing the development of the atomistic world picture.

## Present Shape of the Atomistic Worldview

Certain concepts were formed pre-scientifically from sensory qualities: a body and its position, weight, density, solidity, color, and warmth, the concepts of liquid and gas, of mechanical and

---

\* We are not even addressing "solid-state physics" of the last decades.

chemical interactions, and imponderable natural forces such as light, warmth, and electricity. A *naturalistic* world picture wanted to trace back *all* interactions in nature, including those of living beings, to those mentioned above, without resort to "metaphysical" beings and forces. The *materialistic* world picture emphasized the qualities of matter and maintained that everything could be reduced to the conditions of material things as the knowledge about processes in particular circumstances increased. A *mechanical* view of nature, especially, saw the activating force of all sense-perceptible processes in those forces and potentials that were created according to the pattern of Isaac Newton's central forces. It grew as if on its own out of the distinction between the primary and secondary sense qualities that had been in existence for a longer period of time. Scientifically, it gained importance in the *mechanical theory of heat* and its effects.

Independent of these developments of recent times, there existed in ancient Greek times—in a certain sense anticipating all later views—the *atomistic view of nature*. One can say from a somewhat simplified perspective: The atomism of ancient Greece is connected primarily with the mathematical problem of the continuum. It is not primarily *materialistic*.

On the one hand, the problem of the continuum was taken up in the modern age to explain the compressibility of gases from the spaces between their atoms, while, on the other hand (somewhat later), to attach a picture to the simple number relationships of the observed weight proportions of chemical combinations. Along with the mechanical worldview, the *mechanistic atomism* of the nineteenth century was also formed, the results of which (in the kinetic theory of gases) received a strong measure of importance for the further formation of the atomistic nature picture of a mechanistic bent.

Thus, a conviction grew that the sensory qualities and their associated prescientific concepts of nature description were certainly useful aids toward understanding, but by no means were they the description of "reality."

## The Transition to Twentieth-Century Physics

The philosophical wrestling during that same century with the problem of knowledge against that of reality remains unaddressed here, as is the problem of philosophical materialism. We are here dealing only with the consequences that atomism of the nineteenth century has for physics. It offered what seemed to be a firmly constructed system of final components, the arrangement and movement of which, along with the play of forces and energy relationships, should be sufficient *to explain all present and future natural phenomena—at least in principle.*

The atomistic view of nature still rules today, but no longer with the claim to deduce everything from the qualities just mentioned (formerly called primary qualities). Nevertheless, it is still claimed that ponderable bodies are constructed atomistically, although the borders between the individual particles are "indistinct." There are phenomena and experiments that seem to show strikingly that the structure of crystals consists of atoms or ions, according to X-ray analysis or field ion emission microscopy. Here, individual atoms are *portrayed* in their lattices, even though not "seen." From this perspective, an object is either a sense-perceptible body with the corresponding *classical* qualities or an aggregate—not of real but of possible atoms that answer to the means of examination (e.g., X-rays) according to rules that are no longer those of sense-perceptible bodies. Not only are changes in the understanding of physical material bodies called for, but changes are also needed for understanding concepts such as those that were known as "force" or "matter" in the earlier atomism.

Light, which according to the classical view is solely an energy form and a wave phenomenon, proves to be also a discontinuous representation for the new atomism. Its characteristic wave qualities, which formerly had to explain the observed deviations from elementary ray optics (i.e., diffraction and interference), are now only an approximation at a certain level.

We meet here for the first time the concept of complementarity. Since Einstein's theory of photons (1905), light had to be thought of as partly like waves and partly like particles. There were other phenomena in physics prior to 1913 that, in a similar way, demanded the balancing of mutually exclusive pictures, leading to the atomic model of Niels Bohr in 1913. These form the background for Bohr's concept of complementarity (1925), which he developed especially with the difficulties of visualization associated with quantum mechanics in mind.

Just as corpuscular features exist for light, which for a long time was thought of as undulatory, wave-like features were also discovered in electrons, which had been imagined primarily as corpuscular.* The same was found to be true for all raying of particles.**

The investigations of Werner Heisenberg, which followed up the problem of light emission from atoms, led to a general formulation of mechanics in which all canonically conjugated variables are connected through a relationship that limits the determinability of values of the variables. This is Heisenberg's famous *uncertainty principle* (1925–1927). This fundamental relation of quantum physics sets limits within which the classical wave or particle picture is approximately valid.

Bohr's concept consisted of taking hold of the seemingly contradictory pictures as "mutually complementary traits" in the face shown by reality. They are contradictory insofar as the presence of the one excludes the *observability* of the other, but they are complementary since only with both aspects can we grasp the full wealth and variety of phenomena. The preceding description distinguishes both the conceptual meaning of complementarity and the significance of the uncertainty principle. The latter makes statements only about the limits of determinability. Complementarity, on the other hand, deals with the concept of an object of nature that is not

---

\* Davisson and Kunsmann, "The Scattering of low Speed Electrons by Platinum and Magnesium," 1923; the Davisson-Germer Experiment, 1926.

\*\* Thus, experiments with thermal neutrons from reactors lead, since the 1940s, to geometric neutron optics, in which a wavelength of the order of size $10^8$ cm appears with a median energy of 0.01 eV.

simultaneously thinkable as containing all pictorial features of classical physics. According to Max Jammer, complementarity is the deeper of the two concepts (see p. 345).[13]

Of course, such a statement has to be limited at once: To begin with it can only be formulated with the use of pictorial elements or their rudiments from classical physics. Yet we have to use the ideas rendered here in order to complete the current picture of atomism. By tracing the struggle for an understanding of the facts of quantum physics in this century in the following chapter, we will be led to the question of the mutual connection between the observer and the observed or, in other words, between the instruments of the observer and the elementary events.

*We should go into a short contemplation of certain habits of thinking and ask: What expresses itself in this adherence to the nineteenth-century idea of matter? To say it radically—and thereby naturally be open to misunderstanding—it is a weakness of thinking characteristic of our times.*

# Chapter 4

# The New Phenomena

## The Limits of Classical Physics

Limits appear in thermodynamics when one is dealing with essentially irreversible processes, that is, those that are far removed from equilibrium. In this case, the second law of thermodynamics yields only inequalities, and therefore the events are less determinable.

With the discovery of cathode rays and X-rays, phenomena became known that led ultimately to limits of classical physics. The former were discovered as early as 1859 by Julius Plücker. Their essential electrical nature became apparent by the end of the century, when the brilliant experiments of William Crookes around 1879 were ideal. Their corpuscular nature, however, was deduced only theoretically. After all, advancing atomism had brought the idea of raying particles to acceptance relatively early.

Here emerged for the first time those phenomena that led Crookes to speak of a fourth state of matter: *radiating matter*. This is the anticipation of conditions that in only the more recent plasma physics have been realized in their entirety. Development then concentrated more and more on the emergence of new and confusing phenomena during the last decades of the nineteenth century. X-rays and the rays of uranium and radium were discovered independently, and laws were found in the conceptual order of spectral lines that had nothing to do with the classical mechanistic picture.

In this discussion, three main features of spectral analysis are assumed. Light emitted from glowing gases at higher temperatures gives an indication of the substance from which it is emitted. Glowing solid bodies also emit light, but here the spectrum is continuous, not discrete, in contrast to the band and line spectra of gases. It is these latter that are characteristic of the substances. It has to be added that, in addition to the light caused by raising the temperature, there are many other possibilities of light emission, the most important of which is the influence of electrical discharge.

Historically, it was also important that the dark Fraunhofer lines appear in sunlight—specifically, in the continuous background of the solar spectrum. They led to the statement that a gas, in such a condition that the light emitted by it has a line spectrum, can also *absorb* light. The lines that appear as "emission lines" are exactly those that are extinguished from a continuous, background spectrum.

> It serves a purpose to retain the common use of language but also at the appropriate passages to point to the more or less tacitly assumed concepts of contemporary physics, so that we avoid changes in terminology for those phenomena we will present. Only in this way can we remain comprehensible at all in the face of usages that are now deeply rooted in common as well as in scientific usage. The so-called absorption spectrum signifies no more, initially, than an indication that, after the passage through the substance in question, light shows a spectrum in which those parts are missing that otherwise form its "emission spectrum." If one understands emission and absorption in this sense, one does not a priori maintain that light is made of "single components." Yet we know that the mathematical treatment may be carried out in such a way *as if* light were a sum of independent beams.[22]

Ever more exact knowledge of the structure and the conditions of so-called excitation of spectra went along with establishing the correspondence of line spectra with emission spectra. The existence of discrete lines, however, poses a riddle for electrodynamics.

What "mechanism" in the glowing substance causes the emission of a wave of sharp frequency? Since the electromagnetic theory of light had become dominant, one had to try to understand light emission electrodynamically. The realization of electric waves, which had been made manifest by the experiments of Heinrich Hertz, was accompanied by great difficulties in explaining the appearance of the hypothetical waves. The difficulties consist of the fact that there are laws of the emission spectrum involving integer numbers for which no mechanical or electromechanical explanation exists.[22]

Hendrik Lorentz's electron theory derives all reciprocal effects of electric and magnetic field interactions from the assumption of quasi-elastically bound electrical atoms. Max Planck says of this:

> No doubt, therefore, that Maxwell and Hamilton's theory of electricity was in need of completion and improvement in many fundamental aspects. To have carried out such an improvement, as far and as well as it was possible from the point of view of classical theory at all, may be viewed as Lorentz's greatest scientific achievement. (p. 550)[22]

Empirical spectroscopy ordered and classified an incredible amount of line spectra by the end of the first quarter of the twentieth century. Many empirical laws, most important among them the Balmer series, waited for a unified interpretation. Lorentz's theory—an atomistic one as would be expected—could, as was stated, explain only a part of the phenomena, but by no means all of it.

## Phenomenal Atomism

*Discontinuous* single phenomena were observed for the first time with the detailed investigation of alpha rays emitted by radium and other radioactive substances. There were flashes of light in scintillation screens and isolated traces on photographic emulsions and in the Wilson cloud chamber.

It is interesting to have a look at the pertinent events in Max von Laue's historical synopsis (pp. 37ff)[17]:

| | |
|---|---|
| 1871 | Cathode rays, recognized as carriers of negative charge by C. F. Varley |
| 1876 | Electrical deflection of cathode rays, Eugen Goldstein |
| 1886 | Discovery of canal rays, Eugen Goldstein |
| 1892 | Penetrability of matter exhibited by cathode rays, Heinrich Hertz |
| 1893 | Lenard window, dispersion of cathode rays in air, Philipp Lenard |
| 1895 | Discovery of X-rays, W. Roentgen |
| 1896 | [Pieter] Zeeman effect, establishment [*an assumption*—G.U.] of electrons in the atom |
| 1896 | Discovery of radioactivity, Henri Becquerel |
| 1897 | Cathode ray particle recognized as free electron, George Francis FitzGerald, Emil Wiechert, et. al. |
| 1898 | Canal rays recognized as positive atomic carriers of current, Wilhelm Wien |
| 1900 | Quantum theory developed to explain black body radiation, Max Planck |
| 1904–06 | Relativistic dynamics of electron, Lorentz, Albert Einstein, Max Planck |
| 1911 | Rutherford's dispersion formula, atomic model |
| 1911 | Molecular rays, Louis Dunoyer de Segonzac |
| 1912 | Wilson cloud chamber, Charles T. R. Wilson |
| 1912 | Establishment of the wave nature of X-rays, Max von Laue, Walter Friedrich, and Paul Knipping |

Von Laue says:

Since the Lorentz theory of the Zeeman effect for the charge-bearers emitting the spectral lines of the atoms had, at the close of 1896, led to the same ratio of charge-to-mass, the existence of electrons was definitely established after 40 years of effort. This result was confirmed in 1899 by Emil Wiechert, who, by means of electrical oscillations, made direct measurements of the velocity of cathode rays and obtained agreement with the values derived from deflection experiments. From the historical standpoint, it is interesting to note that Planck in 1900 computed the value [of the elementary charge] from his radiation law and the then current radiation measurements [of the distribution of energy of black body radiation]. This value was much higher than the other contemporary determinations and it is now known that it was far superior to them in accuracy. (p. 114)[18]

These comments show that the agreement among theoretical ideas is considered a final *proof* of the existence of these various particles; that is true for contemporary physicists as well. It is no wonder that the actual observation of scintillations from the alpha rays was only the verification of firmly anchored previous ideas.

## THE CONSEQUENCES OF PHENOMENOLOGICAL ATOMISM

Already von Laue called attention to the extraordinarily surprising fact that cathode rays *thought of* as free electrons can penetrate matter (p. 64).[18] We now know what the significance of this fact is: It led in the first decade of the twentieth century to the conviction that matter "in reality" consists of a fathomless void and that the presumably indivisible atoms themselves must at best be imagined with a picture of energy systems with negligibly small centers.

The existence of an elementary electric charge also belongs to phenomenological atomism; only relatively late was it experimentally secured. This existence is not, however, as obviously atomistic as are scintillations. But there is no doubt—secured by careful determinations—that there were no smaller charges observed than the previously mentioned "elementary charges" imagined at the end of the nineteenth century. The originator of the best *method* of determination, Felix Ehrenhaft, believed for a long time that he had discovered "subelectrons." History ignored his experiments, since, meanwhile, progress of theory left no doubt that electricity *had to* be divided into portions of the quantity mentioned.

I have pointed out that the old picture of impenetrable atoms had to change. Now let's enter into this even more precisely.

If one tries to survey the exceedingly extensive literature—in works of research as well as theoretical scientific and philosophical interpretations—one notices that everywhere *one* principle was observed: Once the idea of the constitutive notion of atoms as elementary building blocks, of electrons as atoms of electricity, etc., was established, science adhered to their "existence" with utmost tenacity.

We should go into a short contemplation of certain habits of thinking and ask: What expresses itself in this adherence to the nineteenth-century idea of matter? To say it radically—and thereby naturally be open to misunderstanding—it is the *weakness of thinking* characteristic of our times, which might seem paradoxical. Hasn't thinking in physics during the previously mentioned years understood not only how to grasp the most unexpected phenomena but also how to connect and, in a certain sense, explain complicated effects in ever-newer and astute theories, and, last but not least, make predictions that confirm the physical fecundity of these theories in a most impressive way? Yet the struggle, as shown in the following quotes, shows that the physicist clung to images far into the formation of quantum physics. His mathematical ability rushed ahead of his intuitive visualization.

Indeed, the phenomena demand a new kind of imaging *at the frontier of the sense perceptible world* that does not bind itself to sense images. I will throw light on the demand for a new imaging from another point of view in the concluding part of this book.

To avoid being misunderstood, I must admit that there was some utterly energetic and sense-free thinking at work in the field of pure mathematics with the usual head start over physics of fifty to a hundred years. This, however, applied to purely mathematical objects and in no way to an eventual application in the world of experience. The abstract character of mathematics made possible a later encompassing of seemingly disconnected facts of physics on a higher level. Owing to these circumstances, theoretical physicists can give the impression of individuals who want to dirty their hands as little as possible through contact with the sensory world. The strength of his mathematical ability was often in inverse proportion to his pictorial creativity, as was still available to Michael Faraday, for instance, to a special degree. It is not accidental that Faraday's images had to wait for the mathematical genius of James Clerk Maxwell to be taken up by theoretical physics.

Let's now turn to individual examples. We remember how the transition of the Maxwell-Hertz electrodynamics arrived at the physical "explanation" of the phenomenological quantities—the dielectric constant and the magnetic permeability. Matter had to be substituted in thought with a system of quasi-elastically bound electrical particles. Such auxiliary means were not only a purely mathematical artifice; they were at the same time naïvely believed to describe physical reality.

Here, the power of theoretical pictures came to expression at the same time as what I call the weakness of thinking in physics. Its persistence is caused by the fact that new confirmations are found again and again for those assumptions once they are made; weakness is shown by the fact that, since the 1930s, almost no essential progress with the help of "models" has been made, although dealing with them admittedly became increasingly subtle in the course of the development of quantum physics. To this day, every representation is inconsistent in itself, because we speak as a matter of course about particles and waves while also insisting in appendices that they are not particles and waves.

Certainly, one could argue that modern physicists are, in fact, only using a historical language and, as a matter of course, learned long ago to think in the new categories and look down in quiet contempt on nonspecialists and their ignorance of professional terminology. Surely this is true in part.*

---

\*   Mathematicians have always been in a similar situation. They use expressions that have their meaning only by virtue of basic explanations or definitions, while at the same time most of the constructs of mathematics are elevated above the intuitively accessible foundations.

Yet it is striking that every clarification of the character of a theory of knowledge concerning the basic concepts of quantum physics has to wrestle with the interpretation of the basic uncertainty of phenomena. A word by Max Born may serve to clarify this, in that he articulates how much the physicist depends on the naïve-realistic picture of an objective world. Born says, in referring to his own popular description of the theory of relativity around 1920:

> I described this method, with the aid of which the concept of an objective world can be maintained, as the greatest achievement of physics, not suspecting that we soon would be facing a totally new empirical situation that forces us to drive the critical examination of the concept of an objective world much deeper yet. Here I have used the expression "a new empirical situation" following Niels Bohr, the founder of the modern atomic theory, the deepest thinker in the area of physics. He has coined this expression in order to emphasize that the birth of new and foreign pictures in physics is by no means the result of licentious or even irresponsible speculation in thought, but the strict critical analysis of an enormous, difficult to survey, complex of facts of collective experience. Physicists are not revolutionary, but rather conservative, and only compelling reasons can motivate them to sacrifice a representation, which was well-founded up to that point. In the case of the theory of relativity the necessity was absolutely compelling, even if the material proving it consisted to a large degree of statements that such and such is not the case.... The second revolutionary overturning in physics, the quantum theory, however, rests on an immense gathering of experimental fact that increases daily. (p. 43f)[3]

Moreover, von Laue points to the difficulties in forming intuitive understanding:

> The younger generation of today's physicists that grew up with the image of the nuclear atom and its orbital electrons can hardly picture the impression that the discoveries of penetrability of solid bodies by corpuscular rays made on then contemporary physicists. (p. 11)[17]

## The New Phenomena

Referring to the difficulty of unifying the mutually contradictory pictures, the same author says on page 41:

> One should not say that this difficulty is insurmountable on principle. *In the interpretation of any experiment theory is already contained.* [Hippolyte] Fizeau's experiments concerning the velocity of light in a moving medium, for instance, "proved convincingly" the existence of an ether, as long as one assumed that the speed of light and the speed of the body could be combined additively. When afterward the [Hendrik] Lorentz transformation showed why this kind of combination was unjustified, the whole of the proof collapsed. The images of the wave and the moving particle both rest on a thoroughly empirical basis; the first originates ultimately from the observation of waves of water, the second from thrown pebbles and similar things. It is not impossible that the future will develop an image that combines both into a unity.

Here, a note in von Laue's appendix concludes:

> Certain concepts of pure physics based on experience have failed in the face of new experiences; better concepts are non-existent at this time. This is a situation not unusual in preceding every greater progress in natural science. But these difficulties cannot compel anyone to change his standpoint, whatever it is, as regards his epistemology; even though they point afresh to the importance of epistemological reflections—like every profound question of physics. (p. 367)[17]

Von Laue still feels the absence of "better concepts" intensely. He, however, supposes they are pictures that are plausible to a thinking that is bound to the senses. Exactly the same is said by Planck in his eulogy of H. A. Lorentz, after discussion of the situation that had been created by the Michelson-Morely experiment and that forced Lorentz to accept the hypothesis of George FitzGerald:

> ...according to which every body moving through the ether exhibits a linear contraction in all its parts in the direction of its movement. It is very remarkable and characteristic for

the staunchness with which Lorentz's way of thinking about physics was so rooted in the mechanical view of nature that he staunchly stayed fixed in this explanation, and did not raise the question, which was obvious under such circumstances, of whether the concept "speed of a body" in relationship to the light ether actually makes any sense at all in physics. Even after the question met him in connection with relativity, he still did not take it up. All his life he instead renounced the introduction of the postulate of relativity while accepting the results of the theory as relationships to a certain extent accidentally valid everywhere, instead of dropping the hypothesis of a substantial carrier of light waves and by that the postulate of an absolute system of reference."

According to Max Planck, Hendrik Lorentz was unable to free himself from thinking supported by having a substantial carrier of light waves. Planck himself did not want to take part in the changeover to the "new style" in the face of the results of quantum physics initiated by him. Max Born relates this in his essay "The Conceptual Situation in Physics."[3] He prophesied that such thinking would persist and that a future change would not lead back but instead to something even further from the classical style. Of interest is the antagonism toward the new concept in Born's characterization. According to Born, Planck lauded Erwin Schrödinger as a man who rehabilitated determinism. Albert Einstein, the renewer of the corpuscular conception of light, fought passionately against Born's statistical interpretation of quantum mechanics. Born called them the "simply discontented" and, using a word game, characterized the particle defenders as *T-totallers* [T for *Teilchen*, particle] and the wave defenders as *π-totallers*. He saw violent antipathy against Bohr's sudden "quantum leaps" and relates, conversely, that he himself had been "accused of betraying the spirit of matrix mechanics" because he engaged in Schrödinger's wave mechanics.

It becomes clear in Born's informative chat that he was wrestling with the new forms of thinking more than anything else. As a contemporary and coworker, he could also speak of the motives and dispositions of thinking on the part of the researchers and, in this way, allows us to participate in the tragedy produced by the absence of new *flexible thoughts* instead of contradictory images.

## More about the New Phenomena

After these anticipatory remarks regarding the development of quantum mechanics out of the so-called new phenomena, we now turn again to the failure of classical mechanics. We can say that the "conservative" physicists of the turn of the century were educated to semiconsciously hold a deterministic, spatiotemporal picture of all physical events. The new phenomena freed them from this prejudice. These phenomena can be summarized as *boundary phenomena*. It emerges that the new phenomena appear at such boundaries where "sensory reality" itself ceases, not the *classical concepts* (as one reads about everywhere today). We emphasize here that we mean the phenomena themselves, not theoretical pictures.

Without threatening their philosophical convictions, physicists at the turn of the nineteenth century could imagine their apparatus and the objects of their research partially in continuum terms and partially as structured atomistically. In every case, they placed the thought pictures (precisely what is called classical concepts today) into the world of the senses. One could not conceive of the idea that this world of the senses itself had borders. Such an assertion would have been rejected as being metaphysical.

Even more recently, the acknowledgment of this particular boundary to the world of the senses was most carefully avoided by factions that battled with one another, the western physicists and philosophers, and those active in the East who adhered to dialectical materialism. Over against all the empty talk, be it about transcending materialism or about justifying theoretical materialism, there remains the simple fact: Both sides painstakingly avoided *acknowledging* the boundary, which has become apparent, *as a boundary of sensory reality itself.* The cause is not hard to find. The result of that previously mentioned attitude, "weakness of thinking," led to the earlier naïve atomism.

I stated that precisely the atoms of theoretical atomism were believed to be *elements of reality*. For only as real atoms were they of value to the imagination of the physicist for formulating new questions stimulating research. When the consequences were confirmed, it was naturally taken as a confirmation of the atomistic representations themselves.

## A Concept of Matter and the Boundaries of Sensory Reality

To avoid being caught in sterile criticism of the formations of concepts of others, we must develop a concept of matter of our own that is not subject to the aberrations described.

While during the nineteenth-century mechanical atomism that seemingly led to a satisfactory picture of matter was developed, a concept was worked out in pure mathematics, which one finds mentioned today in abundance: the concept of the *invariant*.

### Matter as an Invariant

The material identity of ice, water, and steam, on the one hand, rests on the unlimited and reversibly reciprocal changeability and, on the other hand, on their chemical equivalence. It is not too much to ask of naïve imagination to think of one and the same "substance" in differing states of manifestation (although the atomic auxiliary images emerge already much earlier on as explanation, for instance, of compressibility of gases). The previously mentioned unity of the substance "water," however, is bound up with a further general property—the preservation of weight. Moreover, this property of physical objects, well known in everyday experience, does not *demand* an abstract concept of matter.

It is true that the chemical decomposability of many substances into (chemical) "elements" contributed further to the conceptual examination of certain characteristic properties of substances conserved during physical or chemical "state" transformations. The *conservation* of properties that are adequately conceptualized is conceptually nothing but an *invariant*. In mathematics, however, it is thought of without a *substantial carrier*. Whether they are *lengths* with movements, *angles* and *ratios* with tranformations of similarity, *areas* or *volumes* with shearings, or finally, *cross-ratios* with projections, without fail, they are conceptually well-defined "quantities" that receive equal values when formed in the same way before and after the transformation. Even *mass,* which on purpose has not been mentioned yet as a scientific measurement of quantity of material things, is no exception.

Of course, a strong suggestive effect caused by the success of Newton's mechanics in astronomy plays into this. Not only did mass as a certain ratio in the laws of acceleration prove itself to be an important invariant for *all* transformations of matter, but it also appears empirically in a totally different connection as an essential coefficient in the law of gravitation.

To estimate the role properly that mass plays as a naïve-realistic measure of matter, its role in gravitation must not be underestimated. It is manifest that the law of gravitation is independent of the physical and chemical nature of bodies involved; this calls for a deeper explanation. This was found in the mechanical atomism of the nineteenth century with its mass of atoms. This was used to first "explain" the fixed and the multiple weight ratios observed with all chemical reactions in terms of combinations of atoms each of which had its own mass. And then, further, to explain the fact that the masses of atoms were very often near *integer multiples* by picturing atoms as conglomerations of hydrogen atoms. Later on, Einstein's *idea* of *equivalency* of weight and inertia, which he interpreted immediately as a manifestation of a deeper essential identity in the "general" theory of relativity, could have administered the death blow to the concept of matter's adhering to mass. Yet, to this day, *matter-energy* is addressed as *the* core of the new physics, under the influence of the suggestion mentioned previously, in that mass is considered only as a form of energy.

It is true that this concept of matter only plays a role for the "visualization" or philosophical *interpretations* that are irrelevant for the theories. But then, when the worldview of physics is spoken about in this or that way, this invariant appears always in the garb of "mass."

Formulated briefly: The concept of matter encompasses the sum of the sense-related properties of bodies and substances as an invariant—not more, not less. However, from this insight we can, conversely, gain a guideline for the conceptual penetration of natural phenomena. If we say that concept of matter comprises the *togetherness* (i.e., simultaneous appearance) of *sense qualities,* then this concept of matter provides the conceptual grasp of the *boundary of perceptible properties*—namely, when the condition of togetherness is not fulfilled.

It might seem strange that, precisely at this point in the discussion, we repeatedly insist on the perceptibility of the phenomena that form the foundation of matter. After all, one does *not* (directly) perceive the humidity of the air with one's senses. With respect to this, it has to be noted that the togetherness of the sensory qualities mentioned above is to be understood as a methodological guideline and not as a formal definition. One of the most important functions of the formation of concepts in physics consists precisely of the gradual broadening of what is given directly through the senses.

With these amplifications, the principle of perceptibility of the things we try to conceptualize must be preserved. For instance, the problem of physical determinism could come about only through an unbridled and unconscious extrapolation beyond all potential perceptibility.

The prescientific concept of matter rests on general, if indistinct, *experience:* If of the "material qualities" of the substances *some few* are given sense perceptibly, then most of the others appear together with them. Scientifically, this formation of concepts is not productive unless it *explains exceptions* convincingly. In the previous example concerning the content of humidity in the air, the non-perceptibility of it is the result of the interaction of many individual circumstances (which need not be enumerated here). It is essential, in this connection, that it is possible at any time to make the quality in question sense perceptible (in our example, the humidity appears as substance with the aid of the dew-point psychrometer).

*Does it belong to the essence of physics to assume that possible phenomena are manifest only materially or energetically? To this day, physicists are inclined to respond positively to this question.*

# Chapter 5

# Foundational Concepts of Quantum Theory According to Blokhintsev

## Quantized Light

How do the conceptual contents of the new physics appear when followed through concretely? An answer to this question can be gained if one goes through a textbook and pays attention to the conceptual development. Dmitri Blokhintsev begins with a discourse of "energy and momentum of light quanta." He briefly discusses as a basic assumption the introduction of *quantized light* by Planck, as we have touched upon it more historically (p. 128).[2] Then the well-known equation $E = h\nu$ in the form $E = \hbar w$ is introduced. Then we use Einstein's idea that, in addition to the energy, momentum ($p$) has to be ascribed to the light quantum, so that $p = E/c$. This leads to the vector equation:

1. $p = \hbar k$

These equations are combined with the laws of conservation of energy and momentum, well known from mechanics and physics in general, and lead to two more equations:

2. $\hbar w + E = \hbar w' + E'$
3. $\hbar k + P = \hbar k' + P$

The essential content of the introductory paragraph is a rejection of picture images that are too concrete in nature on the one hand, and, on the other hand, a prescription for using the equations (1–3).

Thus, setting $w' = 0$, with the consequence that $k' = 0$, signifies the absorption of the quantized light $\hbar w'$, while the emission of the quantized light, $\hbar w'$, is expressed as $w = 0$ (and $k = 0$). If $w$, as well as $w'$, *is* different from $0$, scattering is involved.

This cannot be handled according to the classical wave theory, as is detailed in a later paragraph by Blokhintsev. First, only this remark is found—that even a partial reflection, which can be described very well by the wave theory, is compatible with the new principles when the energy of the quantum is *not* connected with the wave amplitude in obvious contradiction to classical theory. Otherwise, the different amplitudes of reflected and refracted light would, by virtue of quantum principles, have to have different frequencies. This would imply different colors, which is inadmissible since they are not observed. The idea that the quantum might be a localized particle that swims in a wave is also shown to be untenable. The wave is a purely periodic process infinitely extended in space and time, and a localized particle cannot be ascribed to it.

Blokhintsev starts, for understandably didactic reasons, with a pair of equations and shows the incompatibility of their consequences with the earlier ideas. With this, the postulates of conservation of energy and of momentum are the sole guidelines for operation with "light quantum," the new quantity. The equations themselves naturally express certain nonclassical *phenomena*.

The later paragraphs of his book deal consistently first with experimental proof, then with the appearance of atomism in the light of new ideas, and then with Niels Bohr's theory. The elementary quantum theory of radiation is treated as are de Broglie waves and finally the diffraction of microparticles. In every case, the characteristic traits of the new physics are introduced in this way: the mathematical instrument is described, followed by citations of the abundance of observations that can be interpreted by the new formulas. (The new observations, of course, either contradicted the older theory or were completely inexplicable in terms of it.)

It becomes apparent that quantized light, later called photons, has the characteristic parameters of energy and frequency that, upon interaction, increase or decrease by whole number amounts. This is also the expression of certain phenomena: Light radiation can cause photoelectrons that show stepwise graduated energy states. This radiation possesses direction (hence a vector equation) and a measurable pressure effect (hence momentum for quanta). Also, the discussion of the failure of the older ideas is unavoidable as long as the phenomena that can be described in classical terms are imagined in a naïve-realistic way. I have shown this inevitability to be the case in the introductory chapters. We see here a confirmation of my demand that we do without allegedly tangible pseudo-realities. The failure of classical representations is not a description of reality but it effectively says: It is because we forget the conceptual nature of knowledge that the pseudo-realistic elements become absurd.

Take, for example, the experimental verification of the previously mentioned conservation laws described by Blokhintsev. Of course, with the photoelectric effect, the so-called work function of the electron from the metal has to enter the equation. For the formulae, it is unnecessary (though convenient at times) to think of the electrons as "existing" in the metal. Similarly, it was already necessary to deny the light quantum itself a localization in space. Something much the same might be expected for the electron.

Before we move on to discuss atomism in connection with such a textbook presentation, let's turn our attention to the fact that a wavelength appears in another verification of Compton's law of conservation of momentum:

$$\lambda = \frac{\hbar}{m_0 c}$$

Blokhintsev adds, with good reason, the sentence (p. 8): "*Phenomena in which the constant ℏ plays an essential part are called quantum phenomena.*" We are dealing here with the concrete appearance of that border of the world of the senses, which we pointed to in the section "Matter" (p. 25).

The Compton wavelength is a similar boundary in this field, as is the wavelength of light for the transition from ray optics to wave optics.

## Atomism

The modification of atomism becomes very interesting in light of the new theory. Concerning the elementary particles, one has to say, *"Mass, charge, and other properties of all elementary particles of the same species are the same and are unchanging"* (p.8).[2] They can be seen as structureless objects for a great range of phenomena, but this does not exclude "structures" from appearing with certain high energetic processes. However, since the elementary particles by no means exhaust the "atomism of the microworld," this latter world has to be characterized more precisely. This happens in the following: *"Every kind of composite corpuscle is built out of completely unequivocally defined elementary particles,"* and *"the inner conditions of the composite corpuscles are discontinuous"* (p. 9).[2] The well-known Franck-Hertz experiment serves as an example of the latter. A beam of electrons is sent through mercury vapor. First, the current increases along with increasing voltage. However, it diminishes at an energy level of 4.9 eV, because it is said that the mercury atoms change their inner conditions.

This statement meaningful only under the earlier atomic picture. But even without the assumption that mercury vapor consists of atoms, we still reach a phenomenon of discontinuous change of energy during "excitation." This leads to Bohr's theory.

It will become clear that, precisely because he had abandoned the classical representations, Heisenberg progressed decisively during the early stages of formulating quantum theory. He classified the conditions determining the phenomena as mathematical matrices and was thus able to find in their novel laws the possibility of integrating the mentioned limit, $\hbar$, into the theory.

Mathematical intuition has, at this point, overcome pseudo-realities. It should be our endeavor to sift out, not naïve-realistically conceived auxiliary pictures, but the content of the concept, which is hidden in mathematical formulae. The quantities, momentum, and energy of a quantum mentioned above are actually conceptual correlates of the phenomenon of *light at the frontier characterized by the (essential) appearance of $\hbar$.* Therefore, I agree most closely with the physicist who retains the mathematical form of describing the sensory world and does so without the accompanying pictures. Therefore, the fact that a beam of light possesses energy and momentum can be considered apart from whether we picture it as a wave or consisting of photons.

We might ascribe only *possible* quanta to a beam of light. Just as light does not consist of a wave movement for elementary theory but shows only the new phenomena of interference with suitable conditions appropriate to borderline theories, so also does light not consist of photons if one comprehends quanta. And yet, since Planck, valid discontinuities exist under proper borderline conditions. Using the elementary theory, the wavelengths of light characterizes the border; with quantum theory, the size of Planck's constant, $h$, does this. We will encounter this analogy again in the following discussion of particles.

## Bohr's Theory

Blokhintsev, in his description of Bohr's theory, also connects with historical fact, but only insofar as Bohr's postulates result in the conclusion that the energy of an atom can take on only discrete quantum values.

$$E = E_1, E_2 \ldots E_n \ldots E_m$$

The following statement is an important methodological remark. "Modern theory is not, as we will see, in need of such a postulate and regards quite generally the discreteness of states not as an unconditional characteristic of a quantum system. In spite of this, Bohr's postulate is still correct today, since it can be regarded as visualizing an interpretation of experimental facts" (p. 14).[2] The rule indicated by Bohr concerning the frequencies $w_{mn}$ emitted by an atom—

$$\hbar w_{inn} v = E_m - E_n$$

—is, as we know, simply an expression of the empirical facts of spectroscopy (i.e., the Rydberg-Ritz combination principle and the Balmer series).

The further discussion of Blokhintsev serves only to underscore the unresolvable contradiction between the fact of discrete levels of energy and every classical theory. For the sake of our purpose, we pass over his treatment of elementary quantum theory and radiation and turn to wave mechanics.

## DE BROGLIE'S WAVES

After laying aside the original forms of theory, the fundamental thought of de Broglie is reduced to a *reversal* of the correspondence between waves on the one hand and momentum and energy on the other. To a freely moving particle of energy $E$ and momentum $p$ a plane wave

$$\psi(r, t) = C e^{i(wt - k \cdot r)}$$

is ascribed. With this, de Broglie's fundamental equations

$$E = \hbar w$$

$$p = \hbar k$$

are valid.

Naturally, the question about the nature of these waves and also the significance of their amplitude arises. To start with, however, a connection between the *phase velocities* of the various waves, the group velocity, and the velocity of the associated particles can be found. It emerges that the speed of the phases depends on energy and a group of waves therefore suffers dispersion, so that the group velocity is different from that of the individual phases. The result then is that the group velocity is the correlate of the speed of the particle. With that, every stream of particles is correlated to a "wavelength." The wavelength for high particle energies is very small and can be used to examine the structure of micro-particles. Blokhintsev states: "The idea of a connection between the movement of particles and the movement of waves was so foreign to the images familiar in mechanics that it seemed pure fantasy, and only experiments could lead to the acceptance of it as a valuable contribution to science" (p. 27).[2] This led to diffraction experiments of electron rays in crystals.

Not only did classical pictures of atoms play into the historical course of experiments, but even in modern presentations one cannot forego the auxiliary pictures that we have called *pseudo-realities*. Accordingly, a description of experiments by Davisson and Germer and other authors follows in the presentation by Blokhintsev.

## Phenomenological Results

Light exhibits quantum phenomena in correspondence to matter. Alternately, material movements exhibit diffraction effects—effects peculiar to waves—at certain boundaries of sensory perceptibility.

The correspondence between energy and momentum in both cases appears with the characteristic determining factors of the wave *in the same way.*

If one ignores the auxiliary pictures, a mathematical structure remains. The true achievements of Bohr and de Broglie lie in the mathematical intuitions on the one hand and in the free use of the contents of the pictures on the other. Neither Bohr's model nor de Broglie's waves are meant "seriously" in the sense that would have been valid in the nineteenth century. An example may elucidate this.

One learned relatively early in mathematics to connect a clear meaning to a quantity such as $e^{i\psi}$, although imaginary exponents were impossible, even totally senseless, on the basis of the original definitions. The new areas of application and the structure of mathematical relationships involving this symbol are what has provided imaginary powers with a right to exist in mathematics. It is the *same* with the new relationships discussed above. A particle "has" a wave no more than the power $e^{i\psi}$ "possesses" a counting exponent (indicating the number of repeated multiplications).

Modern physicists argue in this way for a justification of their theoretical methods. What, however, will take the place of pseudorealistic pictures? In chapters 9 and 10, I try to show ways to reach some answers.

## Statistical Interpretation

When discussing an interpretation of de Broglie's waves, the classical identification of particles with wave groups has to be overcome. Blokhintsev draws different conclusions about the nature of these waves. First, it is shown that the spatial dimensions of the group increase (owing to the dispersion). Furthermore, it is noted that, with the use of optical slits, every diffracted part would have to represent a certain part of the electron: "In reality, the particle always acts as a whole and, in the apparatus, the whole particle, and by no means only a part of it, is recorded" (p. 33).[2] However, the assumption that waves "arise in a medium that is also formed of particles" is also untenable.

At this point, the well-known experiments are mentioned. With diminishing intensity and thus longer duration of the experiment, the same pattern of interference arises on a photographic plate. Therefore, the interference cannot be the result of an interaction of different particles. It is this behavior of the particle that leads to Born's statistical interpretation. "According to it the intensity of the waves at an arbitrary point in space is proportional to the probability of finding a particle at this place" (p. 33).[2] The rest is mathematical elaboration for the calculation of the probability of the presence of a particle, with the use of the principle of superposition.

Frequently in physics, the superposition of states is a way of thinking that is as obvious as it is harmless. For instance, one establishes that the temperature and electrical charge of a macroscopic body have nothing to do with each other and nothing to do with its weight. The named qualities can be combined without risking an interaction, at least in first approximation. This has the result, for example, that energy components might simply be added together. In nineteenth-century energetics, such pictures played an important role. One can say that long passages in the famous prizewinning paper by Planck[21] on the principle of the conservation of energy are nothing more than the consistent application of skillful partitions and superpositions of energies.

In quantum mechanics, the same principle enters as a *mathematical* one. With linear differential equations, one knows the fundamental quality of solutions—namely, that their arbitrary superposition yields other solutions. The functions with which de Broglie's waves are described show the same mathematical quality. *Physically,* this is expressed as the principle of superposition of wave mechanics. Thus, the corresponding paragraph by Blokhintsev concludes with the statement, "We thus see that every arbitrary state can be viewed as a superposition of de Broglie's waves—i.e., states having the given particle's momentum, $p$." With this, the probability of the particle's momentum can be determined.

With this, I have finished quoting from the discussion in a representative textbook. We will encounter it again in our study of the equations of motion of a particle.

## The Waves of the New Mechanics: Structures of Information as Physical Reality

Reflecting on the means of treatment I have traced, we can see that the new way of thinking must be aimed predominantly *against* conventional ways of picturing. *How* we have to deal with the light quantum and the mutual pairing of waves and particles results mainly from the fundamental equations. These, however, are idealized expressions of certain phenomenological quantities. Thus, energy and momentum of a particle are placed in correspondence with the frequency of a wave whose amplitude represents not just the intensity of the quantities in question and not just the number of particles, but also the probability of encountering the particles in an experiment.

Let's try to express what de Broglie's waves mean in their mathematical structure. They indicate the total pattern into which the single events will fall. Physics had no concept available for such an organizing principle. The *field*, which also represents a "force-filled space," would be closest to such a picture. However, physicists had learned to think of the fields of gravitation and electromagnetism as *force structures* in the conceptual mastery of the problem of the continuity of action at a distance. They are—also considering all the reservations to be brought forth against a concept that is too naïve—partly effective causes (as electric forces, for example) and partly physical entities (spatial energy distributions). In both cases, however, they could be thought of as quasi-realistic without leading to contradictions. In the form given to them by quantum physics, however, the waves do not have the character of Faraday's fields. Their sudden changes (quantum leaps) correspond much more closely to the sudden changes of a probability system, as when by acquiring knowledge of a certain event the probabilities change all at once. We have not yet learned enough to grant physical reality to such systems of information. However, we are led to these thoughts if we want to state more than just what the waves *are not*.

Why should the conditions for certain possible phenomena be capable of manifesting only materially? This question has been put incorrectly on purpose. Physicists will reply to it. We know immaterial conditions for large classes of phenomena in the fields of classical theory, too, but these fields propagate according to specific laws. They carry energy. In short, they have quasi-material existence. Therefore, put more sharply, does it belong to the essence of physics to assume that possible phenomena are manifest only materially or energetically? Even today, physicists are inclined to respond positively to this question. Exceptions are admitted only within the framework of certain undetermined situations. We should be able to recognize a new kind of physical reality here—maybe of informational character—in the $\psi$ functions and similar means of description.

As physicists of the younger generation, those who have grown into the point of view of quantum physics will perhaps not find anything special in such a train of thought. They can say that one was motivated by facts that are at least as solidly grounded as those of the earlier physics that lead to the development of new methods and have proven overwhelmingly fertile. What I am attempting to do here is not to contradict this physical point of view but to think through its consequences further than is deemed necessary or justified today.

Such further consequences can be seen if we radically free ourselves from the picture that sensory phenomena are results of averages from the sums of numerous "elementary" processes. Consistent with the deliberations of chapter 1, reality is formed only by the phenomena combined with the concepts appropriate to them. Wherever the so-called elementary processes become phenomena, part of their reality causes us to speak sometimes about *frequency* (a wave) and at other times about *momentum* (the corresponding particle). The essence of light is, therefore, not captured by picturing a swarm of photons any more than by picturing a wave. We will undertake similar deliberations after observations on the laws governing particle motion.

## The Laws of Motion of Particles

The main difference between the quantum mechanical description of the movement of particles and the classical mechanical description can be seen in the following: In the mechanics of macroscopic bodies, besides the mass and the position of the body there is only the "state of motion." This state embraces the energy and momentum of the body. In the formal exposition of mechanics using Joseph-Louis Lagrange's equations, and especially through the Hamilton-Jacobi equation of motion, physicists learned to grasp the state of motion of a complicated mechanical system with the categories of momentum and energy. The determining elements with such systems for position and speed become "generalized coordinates and momenta." In a magnificent generalization, Newton's laws of motion are valid for the quantities obtained in this way. For this purpose, it is not necessary to determine the forces in detail if one knows only the functional expression of energy, depending on position and momentum coordinates. This is the Hamilton function that condenses everything into a single formula and allows one to determine all individual movements of the complicated system in the course of time. These movements are therefore potentially given.

In quantum mechanics, knowledge of the Hamilton function of a particle moving in a field, for instance, is sufficient for a physical description of its behavior. But instead of obtaining a differential equation (or a system of equations) for the coordinates through certain differential operations, one forms the *Hamilton* operator. In the simplest case, an eigenfunction results. This provides a discrete sequence of single solutions, or eigenvalues, from which the general solution in the presence of external forces can also be formed with the aid of the principle of superposition. One does not gain descriptions of movement taking place not in time but in so-called states.

## Chapter 6

# The Concept of Probability

## Historical Notes

A discussion on the intuitive concept of mathematical probability and of its axiomatization is indispensable in connection with the concepts of physics.

Mathematical probability arose in connection with the theory of gambling. Gamblers developed at an early stage an intuitive gauge for the "chances" of certain combinations in throwing dice. A question put to Pascal by Chevalier de Mere marks the hour of the birth of the concept. It was customary in his time to make bets on throwing a die: in at least one of four throws does a six appear, or not. If one changed the rule, so that with two dice the bet was made on the appearance of a double six—and also allowed six times as many throws—then the chances were found to be somewhat different. In both cases, they lay near 1:1 and the probability, therefore, nearly one half.

One can imagine that some experience was needed to perceive such differences. Pascal's answer was that the probability of obtaining no six at one throw is $5/6$ and for the next one also $5/6$, so that the probability of not obtaining a six twice in a row is $(5/6)^2$. Continuing, the probability of not obtaining a six in four throws is calculated as $(5/6)^4 = 625/1296$. The probability opposite to that concerns the chances of obtaining at least one six in four throws. Naturally, it amounts to $1-(1/2)^4 = 671/1296$. Thus, it is indeed more probable by a slight margin that, in four throws with one die, there does appear a six than that it does not appear. The preceding reasoning also cleared the path to deal with the more complicated case. As one can *see*, right away with two dice there are 36 possible throws, so that the probability for the non-appearance of the double six is $35/36$; for the non-appearance of the double six in 24 throws, it is therefore $(35/36)^{24}$. The result is 0.509, somewhat larger than $1/2$, whereas $(1/2)^4 = 0.482$. The opposite probability (i.e., for the emergence of the double six in 24 throws) is then $1 - 0.509 = 0.491$; it is *smaller* than $1/2$, while in the simpler game $671/1296$ emerged larger than $1/2$.

As clear as the thoughts providing the basis for this might seem, they very much demand a penetrating investigation of the instinctive elements that played a role in their formation. We merely note here that one tacitly assumes that every successive throw is quite independent of the result of the preceding one, so that it makes sense that the chances combine themselves as described (according to the so-called axiom of multiplication of probability theory).

In the following discussions, we are not so concerned with the mathematical side, with the proper definition of the concept, with the foundation of the rules, or even with their application, though all this will have to be touched upon. We are concerned—in accordance with the general tenor of our discussions—with the experience of the content of those concepts under consideration.

## Comparison with Geometry

In *geometry*, one experiences most clearly the difference between the intuitive understanding and its mathematical form. Theorems of elementary geometry are visualized concretely. Their proofs are both intuitively evident and conceptual at the same time. One *can* grasp them purely conceptually. Then one arrives at the axiomatic structure that was striven for in antiquity. If this is followed through to its conclusion, one obtains a system of implicit definitions that are also axioms. That means that, regarding their content, the axioms are true for one's immediate geometric perception. But for the axiomatic structure, this is irrelevant. It is, however, important that every geometric theorem can be derived purely logically from the axioms, without the content of the axioms playing a role.

The spiritual-scientific significance of the development of axiomatization in mathematics does not need to be entered into in detail here. It may be noted, however, that one of the strongest impulses for the development of pure thinking arose directly from geometry and its strict methods. Thus, the expression *more geometria* referred to the strict form of progression according to hypothesis, assertion, and proof, which originated with Euclid. One of Euclid's axioms gave the impetus toward the development of non-Euclidean geometry. Its content was never doubted, but its form gave rise to the suspicion of its being independent from the other axioms. This geometry led then in a straight line to the axiomatics of today.

Upon axiomatizing a branch of mathematics, one does not ask for the truth of the content of the basic assumptions, but rather asks if they are compatible with one another and what can be inferred from them (and what cannot).

## Intuitive Basic Assumptions

Probability calculations have also been axiomatized. There are various axiom systems, but which are the intuitive concepts? How have they been formed? What significance do they possess for physics specifically? No doubt, everyone hearing the common phrase "the odds for heads or tails when flipping a coin are 1:1" connects the notion with it that this is, first of all, naturally true for a symmetrical coin and, second, that after many flips presumably half will show heads and half tails. A similar situation is true whenever the term chance or probability is used. The *probability is* $1/2$ in the case of coin flipping. For the probability of throwing a specific number using a "fair" die, the value $1/6$ "naturally" is associated. Naturally, because it is intuitively clear that six equally likely probabilities are concerned.

We assume instinctively, therefore, that one recognizes the cases in which certain probabilities are equal. Furthermore, one tacitly introduces a kind of normalization: The probability values lie between 0 and 1—that is, 0 for impossibility and 1 for certainty.

Does this naïve concept, *probability*, somehow indicate a measurable degree of possibility or is it only a measure of the degree of our conviction? Would we not also speak of the probability of a result when we are convinced that an event has taken place but is still unknown to us? In the latter case, one will weigh the probability of the existing but yet unknown result by sensible balancing of all circumstances. An example clarifies the matter under discussion: One has requested a long-distance phone connection from the operator. When soon afterward the phone rings, one expects *with high probability* not just any call but one from the operator. The degree with which this can be conjectured depends very much, as a matter of fact, on how frequently phone calls come in.

## The Subjective Perspective

The opinion that the concept of probability only reveals something about the degree of our knowledge is quite understandable. Nothing but "a reasonable degree of belief," according to this view, can be taken for probability. Perhaps one argues in the following way: The events are determined in and of themselves; the determining factors are merely not known to me. Regarding the changeability of the determining circumstances under consideration, I have to weigh—subjectively—what degree of belief I want to choose and let myself be led there by mathematical deliberations.

Those who want to speak only of *propositions* take a further step. They say: We can take note of events in a way that becomes significant humanly and scientifically only if it can be brought in the form of a proposition. Likewise, our thoughts are meaningful only then—when we bring them to the form of an assertion.

Thus, for example, Harold Jeffreys states:

# The Concept of Probability

> In any case alternative hypotheses are open to the same objection, on the one hand, that they depend on a wish to have a wholly deductive system and to avoid the explicit statement of the fact that scientific inferences are not certain, or on the other, that the statement that there is a most probable alternative on given data may curtail their freedom to believe another when they find it more pleasant. I think that these reasons account for all the apparent differences, but they are not fundamental. Even if people disagree about which is the more probable alternative, they agree that the comparison has a meaning. We shall assume that this is right. The meaning, however, is not a statement about the external world; it is a relation of inductive logic. Our primitive notion, then, is that of the relation "given $p$, $q$ is more probable than $r$," where $p$, $q$, and $r$ are three propositions. If this is satisfied in a particular instance, we say that $r$ is less probable than $q$, given $p$; this is the definition of less probable. (p. 16)[14]

Since one can indeed bring all conceptual relationships into the form of statements, it is possible to limit the meaning of probability to the subjective. That is clear. That controversies arise here becomes obvious from the following quotes. Lindsay and Margenau say:

> Although the usefulness of the probability calculus is apparent in its many applications to physical problems, it is far from easy to fix its exact epistemological status or to justify its applications. In discussing the foundation of the use of probabilities in science we are entering a controversial field, a field indeed in which disputes rage hotly and opinions are nearly evenly divided. (p. 159)[19]

And in 1954, Hans Richter says:

> On the one hand, we stand before a variety of views that already differ in their epistemological meaning of the probability concept. These views accordingly introduce the mathematical model for this concept in a totally contradictory way. Moreover, they are not in agreement as regards to the quantity of objects that are to be encompassed by postulating probability and finally end up in a seemingly futile and sorry state, often even in polemical discussions, whenever there is talk about the methods

that are intended for making statements in concrete situations about numerical values of the probability to be utilized. (p. 48)[25]

And a little later:

We can sum up in three questions the differences of opinion about the introduction of the concept of probability:

1. Is probability *explicitly* definable from previously known concepts, or is it to be introduced *implicitly* as a new quantity?

2. Which objects are to be labeled with probabilities?

3. Does probability signify an objective quality of nature, or is it a subjective valuation of our statements? (p. 49)[25]

Richter speaks of the measure theoretic system as the *explicit* introduction and about Richard von Mises' theory as *implicit*. Disregarding this antithesis, he says:

Probability is supposed to be a mathematically precise model, which somehow corresponds to the prescientific and vague feeling of the certainty that in any given situation a conceivable resulting situation probably will occur.

Furthermore he mentions the views ranging from the position that only well-defined experiments are to be admitted, all the way up to the one that includes our assertions about nature, even those of a completely general kind. He thinks that it is unimportant when restricting oneself to experiments, whether one wants to conceive of probability as an analysis of the events under preconditions of a prescribed experiment or as an analysis of statements about events. Richter believes that, with this restriction, the third question, whether we are describing an objective quality of nature or are merely undertaking a subjective valuation, becomes unimportant.

To be sure, probability is the mathematical-natural-scientific correlate to the prescientific feeling of certainty as regards the future entering an event. Hans Richter then formulates the main problem:

What kind of structure does a system of probabilities need to possess to be valid as a mathematical model of a prescientific feeling of certainty? Why does such a model have to contain the two main theorems of probability theory specifically?

With this last question, Richter is concerned with the theorems of addition and multiplication in the probability theory. The theorem of addition offers no particular difficulties. The discussions arise with the theorem of multiplication. The problem lies in grasping axiomatically the notion of the total independence of experiments from one another. This leads to the question about the nature of chance.

## Chance and Necessity

From a certain philosophical point of view, which does not coincide with determinism, *all* events are *necessary*. "There is no chance." In addition, there are, however, points of view from which events arise only in relation to certain laws. In as much as an event is determined by the law, it can be thought of as necessary; in as much as the event could change while still obeying the law, it is accidental. The *parabolic shape* of the path of a projectile is necessary with respect to the laws of mechanics (not considering air resistance); the *size and position* of the parabola, however, can be thought of as *arbitrary*, depending on *initial conditions*, without violating the general postulate. Accidental qualities are determined by arbitrary parameters. That a particular path emerges in certain circumstances is certainly necessary, but as far as the general law is concerned, it is accidental.

I hold that *accidental* should not be defined in a way that is absolute. A quality is accidental when it is irrelevant for the law in question. Qualities, characteristics, events, etc., may *only* be called accidental *relative* to laws. If something is only "possible"—that is, if its existence is neither enhanced nor negated by the law in question—then to that degree, and only to that degree, it is accidental.

Experience teaches that, as events or their characteristics change, not all those that are accidental appear equally often, not even over longer periods of time. This poses the question: Are there different degrees of chance?

Historically, this difference in degrees of chance is comprehended by the *probability* of an event. Actually, this form of the question is not yet posed correctly. Before considering it more intimately, using the concept of probability and its measure, we have to find a criterion for the nature of chance itself.

## The Elementary Concept of Probability

The question arises as to why, considering the shaky foundations of the probability concept, the numerical results of probability theory are correct. One explanation attempts to conceive of probability as a limit value of relative frequency. To what extent is this mathematical probability an idealization? To what purpose is the concept of limit value necessary, and is it sufficient?

In Richard von Mises' theory, the answer is given by the concept of the collective: The assumption of a limit frequency is, to be sure, a necessary but not a sufficient condition. The digits of a periodic decimal fraction do possess a limit frequency but do not form a random distribution. Von Mises introduces as a further requirement the indifference of the limit frequency to arbitrary selection out of the total sequence. He uses the principle of the impossibility of a reliable gambling system in the same way as the impossibility of perpetual motion machines are used in thermodynamics.

One of the main objections to using the limiting value is that it is not meaningful unless there is an infinite number of characteristic events; thus, it can in fact never be realized in practice. Another objection is that the indifference with regard to selections can never be determined.

One will have to be careful to ensure that the concepts in question that are created freely by thinking, as are all concepts, are consistent among themselves. However, their practical application is a different matter. For example, one would rather speak of a die as not being "proper," when it does not show each number of "spots" one-sixth of the time during a long series of throws, than to assume that the series were still too short. Whatever does not satisfy the postulate is quite certainly not a proper example.

A further difficulty lies in the so-called law of large numbers; one believes that it has been shown that the limit frequency is the result of an infinite number of experiments. This, however, is one of the preconceptions! In reality something different is being shown. The question is: What is being shown?

## Elementary Calculus of Probability

Here, the concept of probability is presupposed in the following form, as Henri Poincaré indicates: It is assumed with dice, drawing lots in turns, roulette, and so on we know a priori that there are equally probable cases. The usual definition given by Laplace—that "Probability is the relationship of the favorable to the possible among *equally possible* cases"—has to be called, according to Poincaré, a relationship of the favorable to the possible among *equally probable* cases. *The results of elementary calculus of probability are nothing but mathematical transformations through which other probabilities can be deduced from those that are known or assumed.*

An example serves to clarify; the two basic rules of the calculus of probability state that when properties are *simultaneous,* probabilities are multiplied, *otherwise* they are added. Those that are simultaneous *are independent* properties; in the second *are mutually exclusive.*

The derivation is generally presented in this way: Select an arbitrary collection of the outcomes. Distribute the two properties, I and II, among them arbitrarily. Assume that:

a) for independent properties, the appearance or nonappearance of one property has no influence on the appearance of the other;

b) for mutually exclusive properties, only I or II but not both can appear for a given outcome.

An example of a) is the property of obtaining a 1 on each die when two dice are rolled. An example of b) is the property of obtaining or not obtaining a 1 on a single die.

Among the $n$ elements of an arbitrary sample—approximately $np_i$ or $np_2$—elements of the one or the other property exist, where $p_i$ and $p_2$ are the probabilities of the properties. The greater the sample, the better the approximation. When events are independent, there will be within the sample with the property $M_1$ a fraction $np_1 \times p_2$ that also has the characteristic $M_2$. The relative frequency therefore tends toward:

$$\frac{np_1 \times np_2}{n} = p_1 p_2$$

For mutually exclusive characteristics, we have correspondingly:

$$np_1 + np_2 = n(p_1 + p_2).$$

One sees that this derivation rests on the following hypotheses:

1. All the events in question have a well-defined limit frequency.
2. Statement 1 is valid for arbitrarily chosen samples.

These hypotheses were the starting point for von Mises' axioms that made use of the concept of a *collective*.

Whatever our statistical suppositions might be, it is necessary to be able to select at random, and it is necessary that the relative frequency of outcomes having the property $M_2$ among those already known to have $M_1$ approximates $p_2$ with increasing precision.

Can this supposition be justified?

*The Concept of Probability*

## Connection between the Logical Concept of the "Accidental" and the Mathematical Concept of Probability

I have stated that the accidental nature of an event is merely its lack of relationship to a system of laws. At the outset, we can speak neither of absolute chance nor of measurable probability. However, through certain known conditions, mainly of symmetry, with dice, drawing lots, and roulette wheels, equal probability can largely be made real to a large degree. These results can be examined by comparing conclusions resulting from equal probability with experience. The various statistical testing procedures rest on this comparison.

Naïve argumentation of probability calculation largely used the principle of sufficient reason: If the initial conditions are made sufficiently symmetrical, *there is no reason* that one or the other event should appear more often; a priori, they are equally possible or equally probable. Specifically, the elementary probability calculation consists of deducing other probabilities from the given a priori ones.

The application of a probability calculation rests in part on the fact that we have imperfect knowledge of the initial conditions (subjective probability). The unknown conditions are then *assumed* to be equally probable, but the experiments with dice inform us unequivocally of the existence of *objective probability*. It is a fact that a random distribution is realized in a particular case; the distribution has nothing to do with our knowledge or ignorance of the conditions of the individual event, although such realization happens with greater or lesser precision, as with any empirical measurement. Since random distributions are *facts*, chance itself is *not exclusively* a relative concept. However, our concept definition states that being random or accidental is also not absolute but relates to a certain system of laws.

An initial consequence is that, for practical application, one has to determine with sufficiently close approximation whether a given distribution is random. If it is, then the results arising from the theory are valid. Wherever a priori probabilities are concerned, the test is needed to determine whether the application of the theory is admissible—that is, whether the a priori probabilities may be accepted as such.

## Forming the Concept of Probability

It seems that one cannot proceed at this point. That is true if we neglect to note the role of thinking in forming concepts. Some of the arguments against the definition by Ludwig von Mises, with the help of the *collective,* completely misunderstand that here we are concerned simply with a mathematical idealization. His *postulates* describe nothing but the immediate picturing of a series of continuable events as we know them from gambling and other "random" processes. I am not speaking now of the factual arguments against the mathematically sound arguments of his requirement for the existence of limit values. Here we are justified to present the necessary correction of von Mises' concept of the collective using measure theory. Both the measure-theoretical and the axiomatic formulation presuppose that one knows from the definition of the concept how the concepts are to be applied practically. The difficulties, which already appear with a concrete assignment of number values in the single instance, show that confidence has already been lost in the intuitive content of the dual role of the concept of probability.

## Overcoming Determinism

The concept of "chance-accidental" coincides with our partial ignorance only in unjustified picturing of a thoroughly predetermined world (previously discussed), because "the event in itself" is thought of as "definitely determined." We do not overcome determinism by falling into the opposite and thinking of some sphere of the world of phenomena as being totally undetermined. Rather, it is sufficient to recognize that determinism is the boundless generalization of a view that is well justified in limited areas.

## Mixing Characteristics

Let's begin with an example. With the game of roulette, the equal distribution of resulting numbers can be understood by the fact that a small indeterminacy in the initial rotational impulse is enlarged by virtue of the great number of wheel rotations to such a degree that the final position of the wheel is no longer continuously dependent on the initial conditions.

The mathematical proof by Henri Poincaré (compare p. 202[24]) is of no interest here. It is predominantly important to enter into the essence of such a *process of mixing* of properties. The postulates of probability calculation require nothing but that the properties of the observed variety of events are "mixed well."

One can well imagine that the limiting frequencies required by von Mises arise from the permutations of finite stochastic sequences, if one imagines the sequences growing ever longer. This thought can be made precise according to measure theory. The notion of mixing the characteristics also makes comprehensible the emergence of pseudo-random sequences. Now, if the number sequence of the decimal fraction of the number $n$ satisfies all statistical requirements of a random sequence, then the equal distribution of the figures certainly does not point to a random process in the calculation. Rather, by the transferral into the limited area of decimals, a law that cannot be expressed by periodicity is blurred by mixing.

*The notion of mixing reaches beyond the subjective and the objective aspects of the stochastic series.*

## The Law of Large Numbers

The problems in connection with the application of statistics are closely tied to the so-called law of large numbers. This law is often formulated in popular discussions in such a way that the only result is the expectation of a limit frequency. But then, only what has been presupposed is proven. Jeffries restates astutely the primary concern here. He writes:

> This theorem was given by Jacob Bernoulli in the *Ars conjectandi* (1713). It is sometimes known as the law of averages or the law of large numbers. It is an important theorem, though it has often been misinterpreted. We must notice that it does not prove that the ratio of $n$ [of the favorable to the possible cases] *will* approach the limit $x$ [probability] when $n$ tends to infinity. It proves that, subject to the probability at every trial remaining the same, however many trials we make, and whatever the results of previous trials, we *may reasonably expect* that $l/n - x$ will lie within any specified range about $0$ for any *particular* value of $n$ greater than some assignable one depending on this range. The larger $n$ is, the more closely will this probability approach certainty, tending to $1$ in the limit. (p. 52)[14]

And somewhat later,

> Therefore, it can be expected with an arbitrary move toward certainty that the series [of partial limit values] approaches the limit $x$ under the conditions of random tests. This, however, is still a theorem of probability and not one that is mathematically proven; the mathematical axiom, that the limit must exist in any case, is wrong because the exceptions that are possible with random tests can be stated.

In more recent probability theory, the law of large numbers is given (in the person of Pafnuty Chebyshev). I am translating the mathematical formulation given, for example, by Boris Gnedenko, into words:

The limit value of probability of the following proposition is for $n \to \infty$ equal to *1*. With increasing *n*, the absolute value of the difference between the arithmetic mean of *n* random variables, pairwise independent, and the arithmetic mean of their expectations becomes smaller than an arbitrary $\varepsilon > 0$. (p. 187)[8]

In other words, it is not that the limit frequency approaches a definite value, but rather that the probability that certain mean values differ arbitrarily little from each other approaches *1*. As a special case, the axiom of Jacob Bernoulli has the form of the probability proposition:

The assertion, that the difference between the relative frequency and the probability of an event with a growing number of cases becomes smaller than an arbitrary $\varepsilon > 0$, has a probability that approaches *1*. (p. 187)[8]

Finally, it may be mentioned that the statement we have formulated is not yet identical to the following one relating to the limiting frequencies proper. The latter is called the Strong Law of Large Numbers (according to Émile Borel, 1909): "The probability that the series of relative frequencies $u/n$ converges toward the value *p* (probability) is *1*." We see that the procedure of implicit definitions is used here.

The main results of probability theory cannot be expressed as mathematical theorems but only as theorems about probabilities. The important factor is that the probabilities *0* and *1* appear. These are no longer *identified* with *impossibility* and *certainty*. To be sure, it is true that impossible and certain events have the probabilities *0* and *1*; conversely, the values of *0* and *1*, respectively, correspond to the predicates *almost never* or *almost always* in the sense of modern analysis.

The difficulty in conceptually grasping randomness lies in the translation of idealized relations into pure mathematical statements.

## Summary and Review

Let's ask ourselves now what the discussion of the various concept formations within the foundations of the theory of probability has taught us. It may seem dissatisfying at first to note that the most important axioms for the applications do not have the character of mathematical postulates but are probability postulates themselves. Boris Gnedenko and Andrey Kolmogorov argue that the limit value axioms I quoted are a kind of superstructure prior to the elementary chapters of the finite problems of a purely mathematical character:

> In reality, however, the actual significance of probability calculation becomes revealed only through the limit value axioms. Yes, without these axioms it is not even possible to understand the real content of the initial concept of this discipline, the concept of probability. The conceivable value of probability calculation consists in that certain large amounts of purely accidental phenomena stipulate in their total effect asymptotically strict, not accidental, [inner] laws. The concept of mathematical *probability* would even then be unproductive, if it did not find its realization in the *frequency* of the appearance of an event with frequent repetitions of an experiment under equal conditions (a realization that is always valid only approximately and never with total certainty) that, however, on principle becomes arbitrarily exact and certain if the number of repetitions is adequately large. (p. 1)[9]

The correctness of this postulate is not doubted in the least when we stress once again that the possible realizations "on principle arbitrarily exact and certain" are not of analytical but of statistical character. In other words, the mathematical establishment of the probability theorem leads again only to probability statements.

Fundamentally, the conditions for epistemological research on the formations of concepts are not any different from those in the field of empirical science on the one hand and those in the purely mathematical disciplines on the other hand. The prescientific experiences

## The Concept of Probability

with frequencies, with empirical limit values, with noticeably stable mass phenomena (as the observations by Pierre-Simon Laplace or Johann Süssmilch), require as much or as little formulation in simplified, idealized concepts as, for example, the experiences that precede the basic concepts of dynamics. No one will think classical physics unjustified on account of the fact that its laws can never be proven "totally exactly" by measurements. The obvious difference between the exact sciences and other disciplines, that in measurements one takes recourse in the calculation of aberrations and that these cannot in the same way be grafted onto another discipline for support, must not mean that the exact sciences are unjustified in any sense. It merely formulates the principles that are the basis for all measurements and brings them into a rational order after the pattern of mathematical physics.

As elsewhere, mathematics fulfills its irreplaceable service by penetrating the mutual dependencies according to quantitative viewpoints. Moreover, it does this by being subject to the viewpoint of cause and effect. One can no more put forward an exact proof for certain statements of probability theory than one can attain a geometrical proof for the exactitude of geometrical measurements. The geometrical exactitude of a construction is of a geometric character; the precision of an axiom of probability must *be* of a statistical character.

Now we may compare the perceivable character of fundamental theorems with the basic experiences from which the theory originates. To master the concept of grasping chance in the accident series, more recent theory of probability has followed this path. First, the relationships observed in the accident series of elementary experience are idealized. Then they are translated into a definite mathematical language. In this language, there are not statements about limit values of relative frequencies, but rather statements are made about probabilities for the discrepancies between calculated and observed frequencies. The limit values of these *probabilities* are 0 or 1.

According to preceding discussion probability is a relative concept. It presupposes an event—whether objectively undefined or undefined only because it is part of a confusing totality for us —in relationship to many other real or possible events. "*Event* is the abbreviating term for all that one wants to turn into an object for quantitative observations of probability. Data of population statistics can be given; empirically speaking, they show great stability in time, for instance. One rightly ascribes to the human being a probability of mortality; one estimates it on the basis of hitherto existing computations (and is prepared to correct the estimation of an objectively conceived value based on refined data). This is a procedure that is not handled any differently by empirical science in any of its fields.

What then is the probability of the mortality of man in population statistics? It is evidently not a matter of concern for the individual human being! Only in as far as man is a member of a population may one ascribe to him mortality, accident, illness, and so on—that is, probabilities. These probabilities are "objective" qualities of the respective objects as members of a *generic species*.

In other connections, I consider that thought very well based that conceives of each individual man as being surrounded by a "probability cloud" on his life's path. This "probability cloud" alters greatly depending on the situations (leisure, work, war, sport) and yet does not touch individual destiny in the least. What is more, according to a beautiful example by Poincaré, this thought keeps the doors open for the intervention of destiny (p. 1386).[23] This idea will occupy us again in the final chapter.

The importance of the *objective* has been overemphasized by Marxist authors, for it corresponds to their rejection of any "idealistic deviation." In spite of that, in this point their thinking is appropriate in the sense of natural-scientific method. Just as appropriate is the subjective conception concerning the mixture of characteristics determined in accordance with theoretical principles, as we discussed.

If the modern mathematical conception views probability as a *measurement* that is defined by appropriate "amounts of events," then this is possibly the best expression of a conceptual construction, mathematically speaking. This construction is equally applicable to each of the different aspects that I have discussed: to the subjective and objective sides, and also to the a priori probabilities emerging from symmetry observation and those disclosed by frequency observation.

This measurement is equally justified as the square measure of geometry, whether in surveying or in Archimedes' calculation of the surface of a sphere. The closest conception to ours comes in the interpretation of probability as propensity.

The mental image of a "probability cloud" surrounding the single individual might be called poetic. Modern physics, however, had to create quite a similar concept for the description of particles.

The emergence of statistics in quantum theory, with the peculiar feature that not only subjective probabilities are concerned as in classical physics, correlates with our conception of probability. Only the *fundamentally* statistical phenomena point to something other than a classically describable physical substratum of the sense-perceptible world. These phenomena are entities on the frontier of the sense-perceptible world and do not bring their nature fully to expression in the phenomena. We will return to this later.

*Einstein could not have chosen a more unfortunate name for the new interpretation of the Lorentz formulas than "theory of relativity"—it's not true that he dealt primarily with relativity. It is closer to the truth to say that he wanted to connect the independence (from the observer's movement) of the speed of light with a principle of relativity...*

## Chapter 7

# The Theory of Relativity and Its Conceptual Constructions

It is characteristic of the necessity to find new forms of thought, beginning with the start of the twentieth century, that compelling forces besides the manifestations of phenomenal atomism and quanta were at work in changing thinking. In relativity, the necessity emerged to widen, to alter, in short, to change the concepts that had heretofore been accepted without discussion.

What distinguishes the essential progress that the theory of relativity has brought to physics? In the way I view this, the question will find a somewhat different answer than that found in popular descriptions. For physicists, perhaps this difference will not be quite as prominent.

### Observations Regarding the Speed of Light

One can argue that it is not appropriate to simply ascribe a speed to the spreading of light. What we actually observe are *changes regarding light*. For these changes, the well-known values of approximately $3 \times 10^{10}$ cm/s have resulted through astronomical and terrestrial methods that are completely independent of one another.

Later a difference was discovered in the dimension of the fundamental units of electrostatics compared with those of electrodynamics: The ratio of the corresponding units not only has the dimensions of velocity but also has magnitude equal to the speed of light (confirmed more exactly by James Clerk Maxwell, 1868–69).

This is a strong support for the electromagnetic theory of light, developed later. From it different ideas were developed to demonstrate the velocity of the earth, using light measurements. Electrical measurements were also supposed to show effects from the speed of the earth. We recall these thoughts not for the sake of a renewed popular presentation, but merely in order to point to the path along which scientific consciousness grew ready to accept Einstein's most radical postulate in 1905. The path leading there is, so to speak, paved with the negative results of all the experiments intended to prove the "absolute" movement of the earth optically or electrically with respect to the light ether.

Augustin-Jean Fresnel suggested that experiments be conducted concerning light's being carried along by moving matter. Hippolyte Fizeau supported the result, at first difficult to comprehend, of only a partial "carrying along of the light" in the moving substance. The prediction of different refractive indices as a consequence of the earth's speed remained just as much without result as did experiments made to prove a directional dependency of the electrostatic attraction of two charges, in relation to their orientation to the earth's velocity. To be sure, Hendrik Lorentz was able to show later that "effects of the first order" are not observable. The search along the few theoretically possible avenues for the discovery of effects of the second order, believed to show an influence on the speed of light by the speed of the earth, was taken up all the more intensively.

Understandably, there existed some head scratching around the turn of the century regarding how the old principle of relativity of movement was to be harmonized with the electrodynamics of moving charges. With respect to this, one must not overlook the fact that, by virtue of Lorentz's electron theory of light, there existed a system of auxiliary ideas. On the one hand, this system allowed an understanding of the unobservability of the effects of the first order; on the other hand, it led to seemingly insoluble contradictions with usual electrodynamics. Along with Poincaré and others, Lorentz reached for coordinate transformations into which time entered like a spatial coordinate. This was a formal artifice when conceived. George FitzGerald then interpreted Lorentz's formulas as a contraction of matter in the direction of movement under the influence of forces that, according to the hitherto existing theory, were to result from absolute movement.

Thus, the ground for Einstein's *axiomatic* postulation was prepared—that light appears to propagate with a unique velocity for every observer. He was able to rely on precise measurements that had been performed by Albert Michelson. These showed convincingly that one certainly could not speak of an effect of the second order. For an understanding of the psychological situation, let's remind ourselves that relativity theory does not rest on the narrow basis of Michelson's experiments, though it often appears to do so. Relativity theory owes its acceptance to a fundamental discrepancy between electrodynamics and mechanical kinematics valid at that time.

## Kinematics as Free Creation of the Human Spirit

Just like mathematics, *kinematics,* the mathematical description of movements, is not pursued as an experimental science. It is created as a free construction in conceptual form, based on instinctive experiences—like geometry. Thus, for example, the elementary rule of the composition of velocities is an immediately obvious procedure for which no one would have demanded an empirical proof, even into our century—regardless of whether it is the famous parallelogram rule or the simple addition of speeds of equal direction. At first glance, therefore, the applicability of kinematics to reality seems as much or as little at stake as the applicability of mathematics altogether.

The merit of Einstein's article "On the Electrodynamics of Moving Bodies"[6] is the fact that he begins from the very point I describe in the preceding chapter, as well as the fact that he does this with the clear consciousness of a need to reshape an entangled situation in electrodynamics. In its essence, the article is the beginning of a *new kinematics*.

Elsewhere in this book, we have mentioned the development of non-Euclidean geometry. *New experiences* were not the decisive factor in this development. It was rather the discussion that occupied two millennia about whether Euclid was right in treating his fifth postulate as an axiom that neither needed proof nor could be proven. The insight into the true axiomatic character of this postulate came with acknowledging as fact that the opposite of this axiom must also be "thinkable" (that is, without inner contradiction). Early on, the thought emerged (with Carl Friedrich Gauss and Nikolai Lobachevsky) that the new geometry could be applicable in other areas of reality. Half a century later, Bernhard Riemann, to whom we owe decisive progress in the field of generalized geometries, clearly formulated the possible physical significance of these general, non-Euclidean goemetries.

With the new geometries, the break with Immanuel Kant's idea that the basic truths of geometry, or at least of arithmetic, are of a priori nature was finally completed in mathematics. The first insights in this direction can be observed in the work of Carl Friedrich Gauss. In physics, however, the pressure of facts was necessary to bring about a similar step with regard to *movement*.

One will, however, do justice to the significance of this step of kinematics if one only frees oneself completely from the confusion created by semi-popular presentations that crept up around this important step of consciousness. A sober discussion is all the more important at this point, because Rudolf Steiner distinctly raised his voice in warning against the "intoxication" with "relativity" in the 1920s. He was especially concerned because *his* voice could be all too easily confused with dilettantish objections, which came perhaps from a good instinct but still wanted to conserve the earlier image of space and time.

Indeed, it is difficult for one's thinking to realize that on the one hand the spatiotemporal foundations of a theory of movement are free creations of the human spirit, but that on the other hand they should not be used outside the area of facts to which they are appropriate. In short, just as much as one can *think* a non-Euclidean geometry, or even more general geometries, *thinking* must be possible also in the field of kinematics. But whether a kinematics is *valid* that deviates from the traditional one (called "Galilean" since Einstein) will be a matter of experience.

## The Theory of Relativity as a Non-Galilean Metric

There is a saying: *Names are destiny.* Einstein could not have chosen a more unfortunate name for the new interpretation of the Lorentz formulas than "theory of relativity"—it's not true that he dealt primarily with relativity. It is closer to the truth to say that he wanted to *connect* the independence (from the observer's movement) of the speed of light with the *principle* of relativity, which not only had its natural roots in the earlier Galilean-Newtonian mechanics but was also not the least bit contentious. Relativity itself was not a shock; the surprise lay in the new proof that Einstein gave for the preexisting Lorentz transformation.

However, it cannot be denied that Albert Einstein made use of the relativity principle to the greatest extent. He was by no means clear (in sense of the word used in this book) about what he was doing; he thought he could deduce everything epistemologically from the relativity principle.

Let's follow his tracks for the moment. In reality (in my view), he wanted and was able to do nothing but connect the observed fact of the independence of the speed of light from the observer's motion with the classical principle of relativity. He stated the latter in such an extended way that he believed he could also let the invariance of the speed of light pop out of the relativity principle. He formulated: The general principle of relativity requires that all natural laws be given a form that makes them independent of the state of motion to which they are related. By this artifice, which was quickly abandoned by more mathematically inclined authors of his time, Einstein believed he had found the epistemological proof for his procedure.

Apart from this shortcoming, which lies in deriving what is an experimental fact from epistemological reasoning, the significance of his encompassing formulation of the new kinematics is not diminished, nor is the personal achievement of its author impaired.

Before we turn to the mechanical consequences, we ask about the special new element of this non-Galilean spatiotemporal metric. It consists, as already mentioned, of the fact that the equations of motion require the inclusion of time, which often is expressed in the way that time has become, so to speak, a fourth coordinate of space.

An important aspect of this new metric is that the invariant that corresponds to *distance* in ordinary space can have spatial *or* temporal significance in the field of the new transformation. Against all differently phrased popularizations, it must be emphasized that time has not *simply* become a kind of fourth spatial coordinate. The difference between space and time has by no means been nullified.

Again, a comparison to non-Euclidean geometry forces itself upon us. Certainly, very many conditions are different in non-Euclidean geometry, but a basic feature of elementary geometry remains: the free movability of bodies. It is also so in the new relativistic metric.\*

Also, in the new relativistic metric there is a definite distinction between space and time, but these two cannot be separated in *the same way* as in Euclidean geometry where one can differentiate between rotations and translations. One could even quite easily call upon the difference between plane and spherical geometry. Within the latter no difference between translation and rotation is known.

---

\*   Of course, we retain the expression *relativistic,* since it became a *technical term* long ago, but it no longer has any relationship to the naïve concept of relativity.

## CRITIQUE OF SIMULTANEITY

Within the framework of physical positivism, conceptual definitions may be made only if they are realizable empirically. If one looks into the criteria for simultaneity, one notices that a concept valid in small distances has, in Galilean relativity, been extended to all space. Whether or not it makes sense to ascribe simultaneity to widely separated events depends solely on the particular phenomenon in question. In electrodynamics we are concerned with a system of relations whose basis is an expansion of disturbances in the field with a fundamental speed. This expansion with a fundamental speed follows inevitably from Maxwell's equations. Thus, one is justified to create a spatiotemporal metric that is appropriate to Maxwell's electrodynamics. This, however, is nothing more than relativistic kinematics—i.e., the Lorentz transformation formulas viewed from a higher vantage point.

We refer to textbook literature for the derivation of these transformation formulas from Maxwell's equations. We can ascertain that the equations of electrodynamics are relativistically invariant (i.e., retain their form under a Lorentz transformation). It can also be shown that the only linear transformations that leave the Maxwell equations invariant are those of Lorentz (also those of Einstein).

One could have arrived at Einstein's theory in this way. As stated earlier, Einstein's personality influenced his entering a path of criticism of absolute space and absolute time and thereby paying homage to a certain overemphasis regarding relativity.

Concerning effects spreading at the speed of light, there is a sharp boundary within the relativistic metric between the events of a temporal nature and those of a spatial nature. That boundary divides the realms of future and past—within which messages have finite speeds—from another realm that is inaccessible in terms of time. We will see from mechanical consequences that it is also plausible to view the limiting speed of electrodynamics, mechanically, as an insurmountable boundary.

Nothing hinders us, however, from conceiving other ways of communication (using nonmaterial means) that have to remain outside the frame of mechanical electrodynamic theories. The justified aspect of Einstein's critique of simultaneity lies in the fact that one cannot infer an actual simultaneity from a naïve conception of spiritual and immediate knowledge, nor can one derive valid proofs in such a way against the new metric.

I know that even a hint of overcoming relativistic limitations made possible by unknown (for now) nonphysical means will be interpreted as a regress into ideas long overcome. Nevertheless, I should state that the limit on the communication of messages, drawn by the theory of relativity, is meaningful only for the physical systems known thus far. I can imagine spaces that are in no way subject to the previously mentioned limitations. To be sure, I have to renounce, as a matter of course, an exclusive dependence on *physical,* mass and energy, relationships. Whether that excludes any kind of information transfer will be shown in the future. To avoid any misunderstandings, it should be noted that one does not now necessarily have to think of extrasensory perception or telepathy (the latter was already mentioned by P. Jordan, p. 318 [16]).

## Length Contraction and Time Dilation

One of the most important and initially confusing deductions from the formulas of the Lorentz transformation is that lengths in the directions of movement seem shortened by a factor depending on speed and that time intervals are longer in the moving system than in the resting system. This statement must immediately be qualified: A moving measuring rod assessed from a resting system appears shortened, while the actual length of the same rod at rest is seen. It is important to note that such a change is not apparent to an observer who is moving, since the measuring rod itself changes, too. The consequences of a "conversation" about this between a "resting" and a "moving" observer have been thought out in fine detail from the start. The transformation equations instruct us that, conversely, to the moving observer the resting system seems shortened by the same factor, with no resulting contradiction. Since at both times the change of length measurement has the same sign, the observers can state only that the measuring rod that is moved *relative* to the individual seems shortened.

In contrast to the previous metric, the results of the measurement depend on the relative movement between the measuring rod and the measured object. One and the same object has a unique length for any observer when the shortening has been taken into account. It follows that length statements are only meaningful along with statements about the frame of reference.

To help clarify a matter that has been thrown into confusion, we point out the difference between the concepts of direction in a plane and direction on a sphere. Directions in plane geometry are determined unequivocally relative to a single coordinate axis arbitrarily taken somewhere in the plane. Statements of direction in spherical geometry make sense only with respect to a definite system of coordinates, such as longitudes and latitudes. Locally, there can be a determination of direction against a reference direction just as in the plane; the comparison of different places brings out the difference between spherical and plane geometry. This analogy is well founded mathematically and extends to the roots of relativistic metrics.

The corresponding fact is also valid for determining time. In contrast to length contraction, we have here time *dilation*, whereby a moving clock appears to be slower than a resting clock. The psychological difficulties here become even greater.

There exists a whole body of literature on the so-called clock paradox. Two observers moving relative to one another set their watches to the same time at the beginning of their journey; the watch of each one is slow in relation to the watch of the other, who considers himself at rest. Each watch has to be behind in time when compared at journey's end. Clearly, this is an impossibility. At the least, common sense can expect that, when their motions are symmetrical, the behavior of watches should be symmetrical.

Even though already strained, the answer of the theory is correct, though a bit flimsy. Strictly speaking, the equations are valid only for non-accelerating motion. A re-encounter with these is impossible. Therefore, the observers depend entirely on communication about the instrument data that accompany them and about the observations of the instruments that are moving relative to them. With that, however, the appearing contradictions can be interpreted away in a way similar to the way those with a length measurement or direction determination in non-Euclidean geometry were. This answer is unsatisfactory, because even the generalized theory retains the time dilation and needs certain additional suppositions to explain the symmetrical case fully.

That these postulates, however, are not entirely without significance can be seen in high-energy physics. Half-life periods of radioactive particles or particles in "excited" states measured in "proper time" appear prolonged when measured by an observer moving relative to them. Certain components of cosmic radiation can only be observed at the surface of the earth, by virtue of the fact that the extremely short-lived conditions appear to be noticeably longer lived on account of their carrier's high velocity.

One must be very careful with the description of the length contractions, which are inherently less difficult to imagine. One must not interpret the formulae in such a way as to reflect the "actual appearance" of the moved figures for the observer at rest. If one wishes to show what a fast-moving system looks like to the observer at rest, one must consider the finite velocity of the propagation of light. What results from this is that a cube moving at sufficient velocity appears rotated and not as a flat disk, as was said for twenty-five years in the literature (*Scientific American*, 1958).

*The Theory of Relativity and Its Conceptual Constructions*

## METRICS IN GENERAL

The metric of projective space offers an excellent example with which one can free oneself from the deeply rooted belief that definite distances exist between points in space and that angles have an objective size. If one knows a kind of metric that deviates from the usual type of measurement, then one can better deal with the relativistic one. But not only that; relativity theory can be understood as non-Euclidean metric of velocities.

In projective geometry, one gives attention only to the so-called relations of incidence and order among its elements. A point is in a line or it is not; a point or a line is in a plane or it is not; two lines in space are either skew or they intersect, possibly at infinity. These are relationships of incidences. Examples of relationships of order are: in a pencil of rays, two lines either separate two others or they do not; in the plane, two points separate two lines whose point of intersection is neither coincident with one of the points nor lies on the connecting line, or they do not separate the lines. Adding continuity, which we do not formulate here, there results a geometry in three-dimensional space that has content through and through—and not just a pitiful rudiment of it. Theorems about conic sections, quadrics, and many other plane and spatial objects can be given using perspective pictures; they encompass an important part of elementary geometry. In this geometry, however, we do not speak about any kind of measure of either distance or angles.

It is quite possible to use a part of the group structure of Euclidean movements (foregoing the characterization of translation) in order to introduce a determination of measurement in a much more encompassing field, namely, that of projective geometry. That does not mean, however, that the measurements that we assign to certain distances and angles would have to be similar to the Euclidean distances and angles that are given intuitively. Yet it is possible to produce an isomorphic picture, for instance, between parts of a sphere and parts of a metricized projective plane in a definite way. In the language of relativity theory this would read in such a way that "an observer" on the sphere and one in that metric would state the same theorems regarding the proportionality of the area of a triangle with what is called spherical excess—i.e., the excess of 180 degrees.

The net result of what we have said for the interpretation of relativistic kinematics is that all talk about measuring rods and clocks is nothing but a visualization of the fundamental isomorphism between equivalent reference systems! The fact that the reference systems, the one judged from the point of view of the other, show quite different measurements is just as little of a surprise as was the case of projective metrics. With that, however, many of the sensational statements of relativity theory have been reduced to their true import. This results from the fact that, at the beginning of the twentieth century, not only geometric but also kinematic thinking freed itself from rigid traditional forms.

## The Metrics of Velocities

By far the most important consequence of modern relativistic comprehension of space and time is the rule for the composition of velocities. Phenomenologically speaking, length contraction and time dilation are initially unobservable theoretical deductions from a change in the transformation equations.*

Another consequence of modern relativistic understanding of space and time is the change that results in the earlier law of the parallelogram of velocity addition. This, however, is nothing but a consequence that results geometrically quite as a matter of course upon the transition from the Euclidean to the non-Euclidean metric.

To describe the heart of the matter sufficiently for comprehension of the concepts being considered, the difference between translations and dilations of Euclidean geometry will be formulated from a projective point of view. Collineations, with a fixed point and a fixed

---

* This modification was meant to bring about the invariance of speed of light with the relativity principle.

straight line, the points of which are all fixed points, are called *perspectivities*. These perspectivities fall into two categories, depending on whether or not the fixed point lies on the fixed straight line. Those in the second, more general, group are called *homologies;* the first kind are called *elations*. All perspectivities with the same fixed straight line form a group of transformations that contain as subgroups those homologies that fix the line and another consisting of the elations. For the non-geometer, any transformation of a figure into a similar figure is called a similarity. Similarities without rotations also contain translations as special cases. The same structure of subgroups arises when everything is mapped in some given way—distorted by perspective, for instance. We have talked about these as homologies and elations. Dilations with similarity are homologies, and the translations are elations when the line at infinity is the fixed line of the transformations.

It becomes evident from this that the elementary free multiplication of the vectors with real numbers signifies the presence of a homology. Furthermore, the remaining operations of vector calculus, such as scalar and vector product, can be expressed in a natural, self-dual means with these geometric operations in the projective sense.

If one now tries to apply the apparatus of homologies and elations in the presence of a non-Euclidean metric, one notices that this is only possible for infinitesimal transformations. One can establish at every point of a general projective metric a tangent parabolic metric that permits ordinary vector addition. In the same way, *Galilean addition of velocity* is valid with low speeds. That is, the Galilean metric is a tangent parabolic metric to the relativistic metric.

In other words, the velocities in relativity theory are "in truth" non-Euclidean geometries. It is not an overstatement to say that the core of so-called special relativity theory is that the "space of velocity" is metricized differently—namely, in a non-Euclidean way and "hyperbolically" in the sense of Nikolai Lobachevsky's geometry. If one advances and sees speed as an intuitively given basic quantity (rarely done in the founding of kinematics) whose dimensions serve for measurement, not for fixing of the concept, it is much easier to think that these quantities of velocity can have their very own rule of composition that lies within the world of experience. The consequences of this for a more refined grasp of space and time are what is paraphrased by the Lorentz-Einstein formulas. Such an introduction to relativity theory can avoid many pseudo-epistemological discussions.

## Relativistic Mechanics

Up to this point, I have stressed in particular the kinematic character of the most important statements of relativity theory. Einstein simultaneously created the generalization of mechanics belonging to it with his new theory of motion. The previously mentioned extremum principles of mechanics were of special use in that. I don't intend to describe relativistic dynamics as abundantly but instead steer directly toward the decisive new concepts.

The most striking and, later, the most important change resulted from the concept of mass. This was for Galilean-Newtonian mechanics an invariant quantity for material systems. Naïve materialism is closely connected with conservation of mass and believes to recognize in this invariant the fundamental property of matter.

In relativistic dynamics, one has to see mass as dependent on speed (in first approximation, a correction term that depends on speed is added). Of special significance is the equivalence of mass and energy, resulting from dynamics, which is expressed in Einstein's famous formula $E = mc^2$ ($c$ = speed of light).

## Mass-Energy Equivalence

At the very moment when it becomes obvious that inertia is no longer a constant independent of the state of motion, the term for the work expended upon acceleration will change. Calculation according to transformation formulas leads to the result that the mathematical structure of the expression for work can be seen as a contribution to inertia. We refer here to the increase of inertia over that of the rest system.

The dynamic increase of inertia proves to be proportional to the work of acceleration. One can call this proportionality an *equivalence,* similar to the way the term is used with the mechanical equivalence of heat. The question pops up whether conversely each form of energy possesses inertia. Again, the mathematical structure of radiation processes suggests a positive answer. Thus, Einstein arrived at the encompassing idea of a general equivalence of mass and energy via the purely mathematical structure of terms for these quantities. Deep down, to him the two entities were essentially identical; mass and energy were only different manifestations of one and the same thing.

By their definitions, however, energy and inertia remain different things. Lindsay and Margenau, among others, point this out (p. 352).[19] Whether one wants to accept the equivalence as a natural fact in the form of a law, or as an expression for identity of essence in a higher sense, remains open here, as in many similar cases. With the formation of concepts for the general theory we will, further on in this discussion, meet a similar situation.

That this was not observed before meets a natural explanation in the extreme size of the mechanical mass equivalence, or vice versa, in the extremely small size of the mass equivalence of a definite energy quantum.

The consequences of this thought have actually never quite been realized. They actually should not make energy more tangible but instead dethrone mass. In any event mass no longer plays the role of the final unchangeable foundation of the material world. These consequences did not need to be elaborated in the early development of relativity theory since common material bodies still possess something like the unchangeable mass of Newton's mechanics in the rest mass. All the more audacious was the idea, already grasped in 1905, of viewing the rest mass as energy densification. This was a concept that was only fully verified experimentally by the "annihilation" processes between heavy particles and their antiparticles, where only rest masses in the form of energy remain.

No doubt, within the limits of relativity theory new formations of concepts will be—one would say—forced on us from out of the mathematical structure, which had to be invented in order to incorporate the mechanics of the new theory of motion. What probably came about through Maxwell for the first time, when he made the electrodynamic fundamental equations symmetric through the displacement current, is continued here in a straightforward way. From Einstein's radical innovations, the path leads on to Paul Dirac's postulates of the double nature of electrons and from there to the general postulate that for every particle there exists an antiparticle.

It is always a decisive element on this path that the mathematical formulas with their structures, of necessity and unnoticed, give to the hitherto existing concepts entirely new content.

Insofar as special relativity theory reaches, there is no direct, immediately observable fact that would lead to this modification of the concepts of mass and energy. One can imagine that, taking the path via the theorem of addition of velocities, the interpretation of collisions between fast particles would have led to the virtual inaccessibility of the speed of light in particle accelerators. It remains to be stated that the historical path, starting from the reconciliation of the invariance of the speed of light with relativity, led to the anticipation of these results.

## Speed of Light as a Limiting Speed

In pure relativistic kinematics, the speed of light can be seen as a practical infinite, because a physical interpretation of velocity addition leads to processes with which the resulting speed is necessarily smaller than the limit speed. Comparison with hyperbolic metrics is also instructive here; by repeatedly producing a given line, the limit of space cannot be reached. In velocity space relativistic addition of speed corresponds precisely to the hyperbolic addition of lines.

Another very weighty argument that the speed of light cannot be attained by material bodies lies in the following circumstance. Upon approaching this speed, the kinetic energy grows by virtue of the relativistic increase of mass beyond all limitations. Thus, there is no imaginable process, for which the theorem of conservation of energy is valid, by which a material body or particle can be accelerated up to or beyond the limit speed. This theorem has remained valid in spite of all modifications of the twentieth century!

Only one principle entered these premises: That the measured value of the speed of light shall be the same in every physical system. If, however, one takes this postulate seriously, then it is clear from the start that fundamental consequences must result for the measurement of space and time. It is true, according to earlier conceptions, that arbitrarily high speeds were thinkable, relative to which light would have to be "poking along."

Indeed, interesting consequences can also be drawn in phenomenological theory for speeds that are not larger than the absolute limit speed but are larger than the speed of light in a given medium. A particle with such "supra-light speed" is then itself a source of light (Cherenkov radiation).

## Synopsis of the Special Theory of Relativity

In looking back one can gather up the main changes in the formation of concepts in key words: critique of simultaneity, elevation of relativity to a universal principle of all nature description, the resulting dependencies of length and time measurement conditioned upon the relative state of motion—i.e., foregoing absolute space and time distances between events. There remain invariant "world" distances (Hermann Minkowski), which possess different space-like and time-like components depending on the reference system, plausibility of a comprehensive mass-energy equivalence, speed of light as being a virtually inaccessible limit speed.

As significant as these changes are, we do not find a truly general relativity in this theory. In it, only those reference systems that are in uniform rectilinear motion relative to one another are equivalent. All popular arguments for a truly general relativity are out of place. Thus, one cannot say that it is only relative whether I move around in relationship to house and world at large or whether I remain at rest and the house with "appended" earth and the sky with its fixed stars moves around me. This purely kinematic relativity is, after all, as old as Copernicus' audacious idea, that, instead of letting the sky with its fixed stars move around the earth, one view this movement as mere appearance caused by a rotation of the earth relative to the solar system and the sky with fixed stars. Special relativity theory has not even a possibility of stating something about rotations.

Nevertheless, Einstein especially fought with greatest intensity for "general relativity" and set it as a goal of the extension of his theory. The pursuit of this thought, however, has led to another result.

## So-called General Relativity Theory

To anticipate the result of a far-reaching adaptation of all formulas to *the postulate of the equivalency of all reference systems,* we can state the following: Far from being equivalent, systems moved in a relatively uniform way are fundamentally different from systems moved in a way that is not uniform. Only fairly new hypotheses about the nature of gravitation led to the mathematical structure that is called, for historical reasons, general theory of relativity and would be better called "covariant description with hypotheses of gravitation."

## The Observer in the Box

Albert Einstein's early main argument for the extension of the theory of relativity lies in an audacious thought experiment: An observer enclosed in a box without any view to the outside could interpret an acceleration given to the box quite well in terms of the occurrence of a homogeneous field of gravity. Indeed, this follows directly from Jean le Rond d'Alembert's principle.

The point of departure is the proportionality between gravitational and inertial mass. In Newtonian physics it was a self-evident natural law that all bodies that show inertia gravitate toward each other. Only the gravitational constant remained to be determined empirically. Einstein asked himself whether there wasn't a hidden sameness of being. He took the extraordinarily precise measurements by Loránd Eötvös as an argument that we deal here not with a fact of nature but with the same thing named differently according to differently defined concepts.

One can turn the same argument in a different direction: A body, falling freely in relation to the earth with its gravitational field, shows no forces between its parts caused by the earth's gravitation—a fact that is well known in the age of space travel as "weightlessness." With the transition to an appropriately accelerated system, the gravitational field can (approximately) be made to disappear. Can all field-like force effects be conceived of as results of a spatiotemporal metric through such appropriate generalized frames of reference? To this question Einstein responded positively with exactly the same daring generalization as in his earlier work. However, his own experiences with mathematical tools taught him otherwise. In order to understand this we will discuss the essence of these methods as far as it is necessary for our purposes.

## The Tensor Calculus

I did not introduce the analogy between the new spatiotemporal metric and non-Euclidean geometry without a reason. In the field of geometry, the transition from the non-Euclidean ones to much more general geometries was made by Bernhard Riemann (1856).

For simplicity's sake, I will state the facts not for the general manifolds or $n$-dimensional spaces but for surfaces. An important theorem of geometry states that on a curved surface the metric can be determined to a large degree through measurements within the region (Gauss 1824). Riemann showed how to master the more complicated relationships for general metrical manifolds in comparison to surfaces. Later elaborations, especially by Elwin Bruno Christoffel and others, supplied the mathematical means for judging the equivalence of arbitrary reference systems. The meaning of all of this is simply the expression of a now much-appreciated geometrical truth: If we described the length measurement of a sphere in a meridian-parallel-circle system, the expression—formulated in another coordinate system—naturally looks different. Certain invariants, however, are characteristic of the sphere in an appropriate coordinate system; plane polar coordinates have a fundamentally different configuration. A part of the elementary nature of the formulas belongs, therefore, to the coordinate system, another part to the object itself. (A balloon, for example, cannot be made from a piece of paper.) The mathematical achievement of the development after Riemann consisted of a flexible calculus that mainly underscores the geometrical invariants of the spatial objects and not the accidental coordinate systems. The developing relativity theory could fall back on this. It is no more correct to say that all space-times are equivalent than it is to say that all spatial forms possess the same inner metric.

After a crisis around 1916, Einstein became clear about the impracticability of a principle of equivalence for all spatiotemporal coordinate systems. From then on something else was the point in question—namely, the necessity of finding adequate mathematical expressions that would let gravitation be derived from metrics.

## Geometrization of Mechanics

The situation of this so-called general theory of relativity can therefore be stated as: "*The program of geometrization of mechanics is practicable.* Movement under the influence of inertia can be calculated in a sufficiently general calculus according to the same process as in geometry. The shortest lines between two points are found to be the geodesics." This more recent theory reveals the deepest basis for the so-called *extremal* principles of mechanics.* The apparent movements of material bodies are geodesic in an appropriately expressed spatiotemporal metric.

Graphically expressed, the Galilean principle of the uniform and rectilinear motion of a body free of forces becomes the straightest and most uniform movement in each spatiotemporal continuum, determined by the momentary distribution of mass conceived according to a geometric pattern. The term *momentary distribution of masses* brings a further peculiarity of the new view to light. In Newtonian theory the bodies followed forces, modified through their own inertia, whereby the forces changed according to the distribution of bodies. Now space and time change with the distribution of bodies, and the movements are—in this new continuum—"force-free," conditioned by the spatiotemporal relationships that change with the movement. This sketchy description justifies our statement: One cannot speak of a general relativity; it does not exist. Instead, there exists an ingenious mathematical structure to portray the effects of bodies on each other quasi-geometrically.

---

\* *Extremal*: The clause in a recursive definition specifying that no items other than those generated by the stated rules fall within the definition—e.g., *1* is an integer; if $n$ is an integer, so is $n + 1$, and nothing else is.

## Concluding Remarks

One can hardly imagine that the quantum innovations in the physics of matter could have been viewed in the same way without the revolution in the basic concepts of space and time arising from the theory of relativity.

Even if the "general" theory is more like a theory of gravitation, the idea to which it gave birth, of the metric field that depends on matter, has contributed much to the conception of radically new mathematical theories, independent of an intuitive comprehension of the quantities entering the theory. In classical physics the concepts were formed through the experience of the senses, or at least according to their model. The unwarranted picturing of atoms that, formed according to the model of sense-perceptible objects, invented an unchangeable carrier for the changing sense qualities, had to be corrected later on account of the new phenomena. However, before this belated "insight" into the untenability of the models that were thought of as realities could expand, the specific as well as the general theory of relativity had prepared the ground.

From the content of the preceding discussions, I raise once more the thought of the four-dimensional distance of two events and its *projection* in space and time, in order to point to a correspondence to it within quantum physics.

We must also remember another fundamental distinction that differentiates the way of thinking only in the theory of relativity and the physics of discontinuous phenomena and caused a profound schism among creative physicists. Naturally, field theory is no longer concerned with coarse, sense-perceptible phenomena. It is now concerned with facts ascertainable in principle with arbitrary exactitude. These facts are only conceivable relative to one another in contrast to the objects of the earlier pre-relativistic conception. Apart from the example of distance already mentioned (and, of course, quantities formed in a familiar way), the parameters (as coordinates, as energy values, as speeds) refer to relativistic objects that have an incomparably greater similarity to everyday experience than do the objects of particle physics.

*I dare make the pointed statement that, if science had accepted Rudolf Steiner's critique of atomism in the wake of energetics of the nineteenth century, an error could have been avoided. Instead, it was corrected thirty-three years later by Heisenberg, who reduced the naïvely realistic interpretations of models to absurdity.*

## Chapter 8

## Concrete Concept Formations

### Classical Laws of Conservation and Matter

Erwin Schrödinger describes the concept of matter at its root: "The concept of a small piece of matter rests on its individuality or sameness." He speaks of a "string" of events that occurs one after the other in time with closest similarity to those following immediately. In our general experience we know how to distinguish the changes in a "string" of events from those that stem from the movement of our own body. According to Schrödinger, this all belongs to the perception of an object, "to that, what it *is* for *us*." The missing pieces of the string (e.g., when we sleep) we supply instinctively. "The considerable persistence of bodily entities is the most significant feature not only of everyday life, but also of scientific experience" (p. 133).[26]

We can say that the prescientific attitude that everyday things retain their identity and can be recognized was transferred to their assumed building blocks. An important support was the idea of the conservation of the quantity of matter. Originally a chemical truth, it was generalized as the conservation of matter itself, extended to physics, transferred to sense-perceptible "reality," and also naïvely ascribed to the building blocks of matter, the atoms.

An example lacking the customary thing-ness is offered by the cloud; it is located in the sky and does yield rain. Its appearance seems to prove its material permanence. Nevertheless, a great deal more may come down in the form of precipitation than the sum of all droplets of which the cloud can possibly consist. The cloud is not a material thing. Here it is easy for us to say what it really is: The cloud is a state into which the moving air enters under the conditions of rain. A candle flame is similar; it has a kind of identity and yet is not a material thing. This was already expressed by the classical statement, "You cannot bathe twice in the same stream" (Heraclitus).

## Matter as State

If we are familiar with Erwin Schrödinger's critique of the particle concept,[26] we should now be able to conceive of the thought: In contrast to the usual way it is perceived, *matter is not thing-like,* physically speaking; it is part of the "real world." In simple language, matter is more a state than a thing. Naturally one will ask: A state of what? In describing particles, physicists had to learn to see them not as thing-like building blocks but, rather, as states of appearance. It is more correct for him not to describe neutrons and protons as fundamentally different things but to conceive of them as states of one and the same nucleon.

The radicalism with which physics itself describes matter as a state of something visibly imperceptible remains unnoticed only because one has kept the terminology "particle," although one has to deny, as Schrödinger says so appropriately, the most important material qualities. "The atom lacks the most primitive characteristics of which we think when we regard some thing in everyday life."

What then are atoms, protons, neutrons, electrons, and the whole host of newer and ever newer "particles," seemingly endless, of which many dozen are known. Are they something whose manifestation is "matter"?

In regard to the flame, we could say that what we see and feel, perhaps even hear, is the result of a process involving gas; its burning is the real process. Light, warmth, and air vibration that we perceive are the results of the same process. Is something similarly true for matter? Are, for instance, the standing waves of electrons in the atomic shells the "process," and the appearance of the impenetrable atom, which sometimes appears recognizable as a particle, merely the result of this process? The answer has to be negative. Even without the electron sheath, the so-called nucleus can appear to have relatively lasting thing-ness under suitable circumstances. In addition, the description of particles through the $\psi$ function is not that of a process in any sens-perceptible or quasi-sense-perceptible substratum, as the earlier descriptions had been. It is only a mathematical description against the physical concreteness of the burning process regarding the flame.

One should say that the "particles" are fragments that appear when one investigates matter at the frontiers of sense reality. To this end, one needs a concept of matter, however, as described earlier.

Whereas, in classical physics, one can follow up on the way these concept formations come about and how concepts already in use can be reshaped, one cannot do this in twentieth-century physics, in which phenomena appear for which no adequate concepts could initially be formed at all. One improvises by retaining the names of previous auxiliary concepts and allowing the actual new meaning to become clear from the use of language in the description.

## Energy

Let's take energy as an example. For classical physics, energy is the mechanical work a physical system can produce. In the theory of heat, energy was known first as a phenomenological quantity proportional to the increase in heat content according to the usual measurement. In the kinetic theory of gases, the energy was then interpreted mechanically as the kinetic energy of the movement of atoms. In electrodynamics, energy first emerged in static fields similar to the potential energy of elementary mechanics. Then the complicated phenomena of fields of currents and moved charges suggested the possibility of a localization of energy in the fields in space.

Around the turn of the eighteenth century to the nineteenth, a difficulty appeared with Planck's quantum theory. Within radiation theory, which had been outlined with some completeness, energy was conceived of through motion of the Poynting vector. Even today, energy remains a useful auxiliary concept; one can see the essence of radiation as energy transport through space. However, if we are concerned with a quantitative description of the interaction between a continuous stream of energy and matter, then an insoluble dilemma emerges that the transition can take place only in minute entities, in quanta. The tangible picture of spatially distributed energy—the movement of which is the cause of radiation effects—has to be jettisoned. In its place, a metamorphosis of the concept of an energy continuum did not take place. Instead, Einstein thought energy is also quantized in the field of radiation (he did this in the exposition of his light-quantum picture). Initially, this radical view had no relation to former pictures. A way that could be reconciled with classical theory had to remain unresolved for a long time. The fecundity of the new hypothesis for the description of experiments regarding ionization through radiation helped further this reconciliation.

Did energy as a physical concept thus suffer a shipwreck? By no means! Ever-new questions about the validity of the law of conservation of energy in quantum phenomena led to the confirmation of the conservation of energy in quantized phenomena to the present day. One can no longer imagine radiation as an energy that fills space in a continuous way. The expansion of the energy theorem by means of the principle of the equivalency of mass and energy is tacitly included. This is compatible with the classical picture of energy through the relativistic expansion of the concepts. Entirely incompatible with the picturing of energy as a continuous distribution, however, is the interpretation of Hertzian waves as "probability amplitudes." After our discussion of the principles, it is clear that classical electrodynamics does not lose its *cognitive value* when it became incompatible with newer theories. In the same sense, the cognitive value of the energy theorem is maintained even though it may be valid only "in the average." Later, I will formulate the insight gained regarding the state of energy.

The interpretation stemming from Born, called *probabilistic,* was introduced initially not for the classical electrical waves but instead for other wave functions that do not themselves describe vibrations of physical things. But what was called the second quantization, that of fields, and is called electrodynamics today, leads to the result just summarized.

What I just did with the aid of the concept of energy takes place in almost every field of the new physics. Increasingly, what matters is not primarily new phenomenological concepts but the invention of mathematical structures that are suitable for the description of the quantitative relationships among measurables by using this or that mathematical operation.

## The Concept of the Quantum Mechanical State

The fact that one can ascribe a state of energy to a physical system is taken for granted in classical physics. In thermodynamics, however, energetically equivalent states with different entropies play a decisive role. In the Hamilton-Jacobi equation of mechanics, one is automatically led to states of energy through the so-called Hamiltonian function. Atomistic interpretation of the laws of gases leads to statistical mechanics. It turns out that for this almost nothing is needed of the intuitive material properties of the atoms. Instead, it is sufficient to speak of the many "identical" mechanical *systems,* whose "states" are represented by points in a multidimensional space called phase space.

The representation of Planck's quantum hypothesis with the help of the subdivision into cells of phase space (an inherently classical expedient) is probably the first step toward the creation of a new concept: quantum mechanical state. That is, nearly two and a half decades before the development of quantum mechanics proper, at least a germ of a concept emerged and would later play a significant role. For the sake of complying with the quantizing of energy transitions in the sense of Planck, it seemed expedient to *quantize* action. This takes place through a division of phase space into volume elements. They might have (for classical theory) an indefinite size that becomes negligibly small in the limit. For quantum theory, however, they must have a definite size.

I emphasize the strange fact that, for Planck's theory, quantum-like or discontinuous *phenomena* did not exist. As is well known, he was led to his hypothesis by the special mathematical structure of the formula. He himself, with great intuition, had conceived this formula to reconcile the energy dispersion in the spectrum of black body radiation according to Rayleigh's theory for low temperatures, with the then newer observations of Lummer and Pringsheim and, at higher temperatures, Kurlbaum and Rubens. To reach a theoretical interpretation of this semi-empirical formula, by calculating the entropy of a system that consisted of many independent oscillators, Planck was forced to assume that each of them was capable of only one set of discrete energy conditions. For that, it was necessary not to introduce a fixed atom of energy but to vary the energy stepwise in such a way that Wien's displacement law resulted. This was indeed achieved when he assumed energy gradations of size $h\nu$.

One can say that rarely has such a fundamental revolution in scientific thought begun so unpretentiously. The consequences were not at that time at all conceivable. First, only some inexplicable peculiarity was the concern—that certain elementary oscillators should not adopt a continuous range of energy levels. No explanation through other classical physical facts could be found, nor has one appeared since then. Strangely, it required a relatively long time and, characteristically, the use of models before the discreteness of spectral lines, well known then, could be interpreted with the help of these energy levels. We will revisit this directly.

Now we will return to the concept of the energy state, which was later made concrete by Niels Bohr. The history of this thought development is generally known: He tried to design a dynamic model in which the electrons in the electrical field of the atom's nucleus undergo astronomical movements, so to speak. Initially, for the hydrogen atom, possible orbits and periods result; however, these are in irreconcilable opposition to a consequence of classical electrodynamics, namely, that the moving charge of the electron loses energy continually through radiation. Niels Bohr had to contend with this as another consideration produced beguiling consequences. For the first time, it was possible to use a simple model to classify the confusing variety of spectral lines. The emerging frequencies correspond to different energy levels of electron orbits. An oscillating electric atomic system, originally suggested by Hendrik Lorentz, allowed new interpretations of spectra, thus opening opportunities that required putting up with "physically impossible" states of the atom. One knows how, after a struggle, Bohr decided to ascribe "radiation-free orbits" to the electrons and view as the cause of the emission of light quanta the leap-like and sporadic transition between orbits.

The frequency of these quanta is no longer determined by the original idea of a real orbital movement of the electron with its period but only through the Planck condition $E = h\nu$. $E$ has to increase from orbit to orbit in simple arithmetic progression to match the frequency steps of the previously indecipherable Balmer series. At this point, Max Planck's abstract requirement suddenly dovetails with the once-mysterious empirical material of spectroscopy, which had for decades oriented itself with Balmer's "harmonical" key. Johann Balmer had arrived at his formulas (empirically verified in such an overwhelming way) using laws of whole numbers in harmonic consideration in the style of Kepler and Neoplatonism.

Meanwhile, the fact that Bohr's model cannot represent an object reality has become clear to physicists. Its usefulness as a visual scaffold, with whose help one can know how to deal with later formalisms, is unaffected. Thus, elementary instruction and popularizations now offer this model as a description of reality.

One cannot simply say that new concepts have been formed. Rather, visualizable characteristics that contradict each other have been bastardized here, under the sole dictatorship of certain formulas that yield predictions of phenomena.

Here Plato's old saying of "saving the appearances" came to be honored anew, even if via the detour of the phenomena's formalized representation at a prominent point in the development of physical imagination.

On the other hand, new ways of viewing nature were struggling for emergence—for instance, Bohr's "complementarity," whose interpretation we will attempt elsewhere.

## Bohr's Model

The energy states, called for by Niels Bohr in his semi-visual model, were taken into the quantum mechanics that developed afterward. Initially they were states of thought that pertained to atoms without sense-perceptible characteristics; this justified the term "state." Later, however, it was only permissible to speak, in dealing with quantum mechanical systems, of "states," ordered according to whole numbers, whose discontinuous transitions took place according to certain laws of probability. The "correspondence principle" of Bohr, utilized in the first phase of development, rendered it possible to concretize the formalism of states and their transition probabilities toward their use in certain physical problems.

If we now ask what this "state" is, we have to admit that it is only a *name* for the fundamental property of a formalism, with which one has learned to quantitatively describe in an exact, systematic way an extraordinarily encompassing abundance of phenomena.

Regarding the concept of mass, Henri Poincaré admitted at the beginning of the twentieth century, "We have thus achieved nothing and our endeavors have been in vain; we have of necessity to return to the following definition, which is merely an admission of our powerlessness: Masses are coefficients that one enters into the calculation for convenience's sake." Thus, we must admit all the more that a concept of the state is merely a name for an important part of a formula.

Do we really have to do that? Isn't there a concept that has not yet penetrated into full consciousness hiding behind the mathematical abstraction? We may try to grasp it in the following way. The energy state turns out to be an eigenvalue in the mathematical formalism of quantum mechanics. This points to the bond with the observable facts (phenomena). The use of the energy function as a Hamiltonian operator shows this to be an energy state and leads to the scheme shown as a diagram:

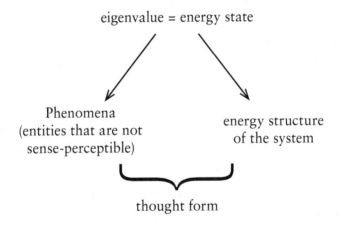

The thought form here is an element of an informational system and not a model-like, comprehensible entity. Consequently, the energy state is the key to the manipulation of the *sense-perceptible aspect* of something that is not sense-perceptible upon its manifestation.

## The Novelty of the World of Particles and Quantum Events

If one surveys the extensive literature contributing to philosophical interpretation of physics in the twentieth century, one finds two distinct characterizations. One holds that the laws of reality into which physics has penetrated are not possible to perceive. The other, in a thousand varieties, is concerned with what the new entities are *not*. Both have profound justification. Let's begin with the second: It must be made clear that the materialistically grasped concept formations, especially of nineteenth-century physics, have to fail if one wants to use them to penetrate that new layer of reality. Until now, this layer is, fact, accessible only to mathematical formalisms, which in a dramatic situation are partially "invented," partially won by continuous elaboration of already existing theories, and partially developed from daring postulates. Examples of these are Bohr's invention of the orbit, the elaboration of phase space, Einstein's daring postulate of light quanta, and taking Bohr's orbits to be radiation free.

Let's use the consequences of our epistemological considerations. These led us to the ideal or conceptual character of many things that were assumed to be just facts of the sensory world. We especially remember the caution we employed while forming the concept of matter. Again and again, we are led to recognize the thoroughly ideal, conceptual character of almost everything that materialism (in physics, *naïve realism*) placed into the outer world as facts without further ado. Nevertheless—and this is the main point—in contrast to current belief, we have acquired the conviction that these elements of reality are more than mere *conceptual shadows* and instead comprise the full reality in connection with the content of sensory perception.

However, we must remain conscious of an important qualification that is currently expressed more or less clearly by all responsible researchers. This qualification is that the world examined by the new physics is *not* the whole world. An outstanding representative of physics, who presents this in a straightforward way, is Walter Heitler.

Materialism in the past included confidence that not only the exact sciences had secure support but the phenomena of life, of emotional behavior, and of conscious reflection did, as well. All were taken to be more complicated and ever-less comprehensible interweaving of fundamental, blindly ruling, iron laws of inanimate nature. This materialism held to the honest conviction that this rule of iron law would result of absolute necessity from the development of exact sciences. That this conviction disappeared from middle European and Western scientists brought the accusation that they were "idealists" by dialectic materialism. This reproach is made even if an ill-concealed agnosticism is called "dealism" through the insistence of dialectical materialists on the merely subjective character of our concepts.

## Theses

With very few exceptions, the previous materialism has merely been negated. As yet, no similar new and modern way of thinking has emerged that would be adequate for the new layer of facts. Here, we highlight sensitive points in the form of theses:

1. Most of the concepts of classical physics are legitimate *insofar as they are phenomenological descriptions.*

2. These concepts must not be extrapolated beyond the validity of the idealizations on which they are based.

3. The descriptions by differential equations of processes occurring in inanimate nature bring to light a certain structure that we find in the phenomenal world.

4. The description by differential equations does not grasp the full reality of the world of substances and forces and their interactions, and it is not really "causal."

5. Determinism can be deduced from such a causal description. Determinism throws up a smokescreen of pseudo-problems arising from illegitimate extrapolation.

6. The boundaries of classical physics can be reached on two paths. One might call them the path through the curtain of sense perceptions (veil of *maya*) and the path of "contemplation."

7. On the first path, we advance into new realms beyond what is sense perceptible when we encounter the

phenomena of atomism, for the things appearing here show unavoidably that they do not possess all the characteristics of material objects.

8. On the second path, contemplation of the connections among mathematical formalisms, which describe the phenomena factually in their various areas, leads thinking into insoluble contradictions unless assumptions of quite a new kind are made.

## The Fundamental Probability Propositions of Quantum Physics

Ever and again in discussions about the proper meaning of concepts in the sphere of quantum phenomena, one encounters disagreements about how the statistical statements in this field are to be understood. Early on, the so-called Copenhagen School decided to view the fundamentally statistical character of the quantum leap for instance but also of other propositions essential to this theory, as a characteristic trait that cannot be avoided. The rejection of the "hidden variable" is based on this view. Other opinions are found in two extreme camps. One of them, headed by Louis de Broglie, is always searching for thought models that are supposed to explain the emergence of merely statistical statements—"explain," meaning from frames of reference that can be determined themselves and are nonstatistical. Sometimes these are called deterministic because their tightly woven laws (which can be expressed in continuous functions) were equated with the total determinacy of their results—in short, continuity of mathematical solution equals physical determinacy.

The other camp uses the Copenhagen interpretation in establishing new boundaries of knowledge. "Quantum physics cannot answer all questions regarding the essence of things but it answers all legitimate questions." Implicit here is that legitimate questions concern only the subjective ordering of phenomena under the restriction of the uncertainty principle.

Werner Heisenberg's analysis, using the then current models, shows that the concept of an elementary state that is totally self-determined is not unthinkable but is in no way *ascertainable*. Above all else, the discussions here teach that totally determined states of classical physics must not be transferred onto a reality that exists in and of itself. They belong to another sphere altogether—the sphere of mathematical idealization. Indeed, it leads to metaphysical constructions if one ascribes these mathematical idealizations to a thing-world formed after the pattern of naïve realism.

I dare make the pointed statement that, if science had accepted Rudolf Steiner's critique[29] of atomism in the wake of energetics of the nineteenth century, an error could have been avoided. Instead, it was corrected thirty-three years later by Heisenberg, who reduced the naïvely realistic interpretations of models to absurdity. However, Heisenberg did not use purely epistemological arguments but rather an idea that had become an axiom by then—that elementary energy transitions have to happen in quanta.

Heisenberg's critique leads the Copenhagen interpretation to subjectivism and the other interpretation to the objectification of thought creations. Here lies the actual root of the problem. The need to observe not only real but also possible interactions, particles, and states presents us with the question about the reality of these elements—which have become so indispensable for the exact description of nature. Western physicists distrust the power of thought: They make overly cautious or vague statements. When discussing the dissolution of matter into systems of mathematical relationships, physicists and philosophers of physics in Eastern European declare matter to be something they *must think* for the sake of a "rational" description of nature.

Here a strange notion emerges. In several regards, much more energetic *thinking* is employed in the East to penetrate the physical world spiritually, only to name the result of such work *matter*.

*Almost all fundamental laws have to be expressed in mathematical form. Only by that do they receive a clear meaning. Conversely, only mathematical formulation enables us to free ourselves from anthropomorphic naïveties.*

## Chapter 9

# The Phenomenology and Mathematics of the New Physics

## Phenomenology in the New Physics

The roots of classical physics are based strictly on phenomena. Their formal evolution introduced "rational description," especially in mechanics. In fact, the great "theories" of mechanics are abstract (i.e., removed from direct observation via chains of "idealizations"), yet they are not necessarily based on hypotheses that use unobservable elements. The mass point is an idealization. It is true that this mass point is not thought of as a concretely observable thing, in that one can trace its movement per second, but it is the idealization of a small body, to which one "naïvely" ascribes an identity through space and time. The propagation of a wave through light years, all the way to unimaginable dilution, is already a borderline case. Being actually an idea of classical physics that received its final definition in Maxwell-Hertz electrodynamics, it is at the same time a hypothesis in electromagnetic optical "theory." Even if one intends to describe the transition to the new physics in a roughly factual way, one cannot proceed *phenomenologically* in the same way as was previously possible. As to why this is the state of affairs, the historically foremost and—as I think—deeper reason is not the "nonclassical" nature of the new phenomena, but something quite different.

It seems that in the leading thinkers at the turn of the nineteenth century new thoughts were born that *anticipated,* with a certain inner consistency, essential features of the new "world picture" (as it is called among physicists and philosophers), despite all the difficulties of picturing it. Planck's quantum of action and Einstein's equivalence of mass and energy are *pure thoughts;* Bohr's quantum jumps and stationary states, Pauli's exclusion principle, Heisenberg's uncertainty relations, and Schrödinger's $\psi$-function are *universal thoughts,* which could not have been thought without *hypothetical pictures,* at least to start with. Dirac's electron and especially his positron as "hole," Fermi's neutrino, and Yukawa's meson are all *picture thoughts* of a completely new realm. But they also all became phenomena in a shockingly real sense.

One cannot help regarding the evolution of the new physics from a different phenomenological perspective as well. The new phenomenon of the theoretical thought experiment emerged alongside the new phenomenon of the experimenter. A new connection between the nature of thought and phenomena announced itself.

There are quantum mechanical analogs of measurements, but the possibility of serious contradiction arises when one tries to identify "objects" that can be defined for themselves with those measurements. These kinds of contradictions are typical: If one thinks in terms of the particle picture, under certain conditions the inherent wave qualities of interference produce dispersions such that no definite tracks may be ascribed to the particles. If we think in terms of the wave picture, then after the quantum leap that may be induced by the wave the disappearance of the wave throughout the whole of space is physically unimaginable. Add to this the fact that, in a quantum mechanical system for electrons and certain other particles, no two of them may be in the same state at the same time. So long as systems of the usual small order of magnitude are concerned, this state of affairs is peculiar and not motivated by any otherwise known phenomena. New difficulties of picturing arise regarding systems that are extended over large lattice plane distances, such as in a crystal. Here it is not physically conceivable how the single units "know" which states are "occupied." By saying that certain "totality principles" are in effect, one merely has another word for it but neither a motivation for nor a derivation of the case in question. One cannot rightly call it a "phenomenon" since it plays only a conceptual role in "states" and "quantum numbers"; yet a state of affairs shows itself, one that governs extremely differentiated phenomena and is empirically secured by numerous confirmations of its consequences.

To conclude: The "new phenomena" take place in *part* in the macroscopic consequences of micro-processes that are inaccessible to material description but also have revealed themselves as *thought phenomena*. This term, coined with reference to the physicist's "thought experiment," points us to the hypothetical pictures on the one hand, which were indispensable heuristically, but on the other hand, also to the fact that a large number of the new phenomena first existed in thought and only later were realized experimentally.

## Excursion into Concepts That Are Grasped... yet Not Grasped

In the dialog, *Theaetetus,* Socrates discussed in practical terms what it means to hold a concept and still be subject to error, with the example of the dovecote. The owner of the dovecote has all doves in the cote, to be sure. However, if the owner wants to grasp a certain dove, one can err and seize the wrong one. It is just like this with a thinker, who "grasps" the concepts, to be sure, but in a given situation seizes the wrong one. Similarly, we can already grasp concepts in a certain sense, and yet not possess them concretely. To use another picture: While hunting lions, we know that a certain lion lies in a thicket; we hear the lion roar, see its track, smell its scent, and yet we do not "concretely" know the lion. It is like this with certain concepts of modern physics; they remain hiding in a thicket of formulas. We know their tracks, using them we can classify a multitude of experiences, and yet we do not "really" hold them.

At this point we refer back to our epistemological chapters. As long as we do not succeed in realizing the content of the concept within a definite fact of experience, we must concede that we only have names for formulas or parts of them. We will make the connections arising from this after the discussion about further concept formations of modern physics.

## The Significance of Mathematical Concept Formations in Classical and Modern Physics

The role of mathematics in classical physics could be thought of as being extremely unimportant in spite of its encompassing apparatus. Its purpose is "merely" the description of quantitative relationships using appropriate abbreviations and the symbolic notation of the conceptual deduction by way of different abbreviations. This restrictive interpretation is entirely appropriate for vast areas of experimental physics. Very quickly one reads otherwise, prompted by textbooks and dissertations. Almost all fundamental laws have to be expressed in mathematical form. Only by that do they receive a clear meaning. Conversely, only mathematical formulation enables us to free ourselves from anthropomorphic naïveties.

Thus, for instance, the proportionality between force and acceleration in Newton's fundamental law does *not* state that acceleration is a "result" of force, as one would be likely to assume at first. If one wants to think at all causally at this point, one can state that the acting force is the cause of the resulting *motion* that arises later, so to speak, as the effect after the cause. Expressed mathematically, *the acceleration is the cause preceding the motion*. The proportionality between force and acceleration expresses the matter more adequately: The acceleration is the kinematic cause, the force, the dynamic basis of the arising motion.

On the other hand, only mathematical formulation permits a fundamental statement to encompass an infinite number of single cases, each with its own peculiarities—for instance, those of projectile parabolas that can be found deductively from rather insignificant premises. This impression becomes stronger, especially in the higher levels of mechanics as well as in thermodynamics. Last but not least, this impression becomes stronger in electrodynamics, to the degree that one feels that the essential element of the corresponding theoretical connection is to be found in the mathematical structures and not in the concepts that have been offered by experimental physics. One merely needs to recall the role of the so-called dimension arguments to recognize the greatly clarifying effect of strict mathematical formulations.

Another aspect of the role of mathematics emerges when one sees it as a means of obtaining correspondingly different results for different kinds of problems. Quite specifically, one sees the significance of the form of Hamilton's dynamic fundamental equation in the fact that the general configuration is of more interest than are the concrete procedures for handling motion.

If we introduce the concept of a *possible event* in classical physics, we do so only in view of the general statement that encompasses all possible special cases. At this level one does not assign to the possible event any definite kind of significance in itself. A glance at the quantum mechanical development shows us that matters are quite different in modern physics. Here we must employ in our calculation all those "virtual" interactions of a photon on its track as possible events that, however, do not become real. In older theories, these were viewed as "real" interactions between the light wave and the particles of the material media through which the wave runs. The result therefore depends on the net of possibilities through which the photon moves, a thought that flies in the face of every idea of classical physics about the cause or determination of a process by accidental circumstances.

Many discussions about the significance of the so-called extremal principles in physics originate from the desire to see, for example, the sway of the divine order or an economy of nature in the specific configuration of the real events versus all the possible ones, through the "smallest force" (Gauss) or the "smallest effect" (Maupertuis). In more recent discussions about this problem, one is usually satisfied with questioning the actual extremal character and claims only that the quantity under consideration is "stationary," a necessary but insufficient condition for the existence of an extremum.

Since just at the transition to relativistic as well as to quantum physics the extremal principles show a point of departure for a completely new formulation of the questions, we had to continue to discuss them here.

In conclusion, we can state that the highest level of classical physics consists of defining the possible *form* for events out of a higher sphere of formlessness and with the aid of mathematical structures. Within the framework of classical physics, only the accidental starting conditions determine which of these possibilities is made real. The idealization of the events under consideration, in the shape of mass points, waves, or fields, is accomplished in such a way that one can imagine the realization with any and every desirable exactitude. Here the main point is to find the transition to modern physics.

We can now return to the endeavor of interpreting the meaning of the mathematical apparatus of modern physics. It is not my concern to acquire *new* interpretations for detailed events. Rather I want to allow for a conversation between the nature of applied mathematics and the human spirit so that mathematics can express something about that realm that it helps us to quantitatively master.

## Equations and Differential Equations

Originally, quantitative nature description was done by simple proportionality, as in Archimedean or Galilean mechanics. One may think of the axiom that, in balance, levers are inversely proportional to the load, or one may recall the statement that the speed achieved in a fall is proportional not to the distance that has been traversed but to the time spent. Truly, the difference between the first postulate, originally by Archimedes, and the second, which we owe to Galileo, expresses much about the change of consciousness within the time span of 1,800 years, but it is immaterial mathematically speaking. The difference lies in the fact that, in the first instance, static conditions are comprehended and, in the second, those of motion.

However, in Galileo's postulate we already have the germ of the next step, taken by Newton, which lets us describe the finished process of motion and a general frame of reference that encompasses all conceivable motions. The reader may have anticipated what I am aiming at: The development of celestial mechanics leads to the description of the world by differential equations.

The differential equation (in the simplest instance) connects the values of a function not only with argument value but also with the changes ("differentiations") in the function itself. Therefore, it is not a concrete functional relationship that is created in the case of a common differential equation; in a certain sense, an "empty form for functions" is characterized. In this same sense the function is similarly valid for differential equations of a higher order and for partial differential equations. The particular solutions can be determined concretely in many different ways—e.g., through initial values or through boundary conditions).

In physics, the particular solution can be determined from the description of boundary conditions; the differential equations describing mechanics and electrical fields are of a second order. This comprises the causal character shown by classical physics, but in a narrow sense. Out of this a nonphysical determinism arose by way of extrapolation. Pascual Jordan stresses the statement "that in the living organism the physical *principle of close range action* (which is the visible expression for the fact that physical laws are partial differential equations) is partially put out of operation." He then acknowledges that such integral laws "represent a very natural and obvious generalization of those physical natural laws that we know." He also points to the variational principles of physics in this connection (p. 129).[16]

## The Differential at the Border of the Imperceptible

Differentials are thought of as crude in mathematics (by now they have become fashionable again in nonstandard analysis); at most the technicians, perhaps even the physicists, have a right to differentials as convenient symbolic aids. Seriously: We do not want to negate the great labor of thought accomplished by Augustin-Louis Cauchy and Karl Weierstrass in making comprehensible statements about the limits from the somewhat nebulous concept of differentials of different order. The incomprehensible element was shifted into the problem of the continuum, which received its magnificent treatment from Cantor. Without the exemplary rigor of the well-known "critique of methods," no certainty could exist in the increasingly complex system of analysis.

Having paid respects to the taboo involving differentials will now turn attention to the frontier of consciousness to which they lead us. By writing down the simplest differential equation, we express the process of coming into existence of quantities and figures in a mathematically accessible form unavailable to intuitive representation. This is not the only place where mathematics has crossed barriers to imagination. Moreover, the introduction of imaginary and complex numbers, ideal numbers in higher arithmetic, and many similar experiences of mathematical objects transcends immediate pictorial representability.

What is peculiar with differential equations, especially the ordinary ones, is that the instantaneous rate of change of the quantity to be determined also depends on this quantity itself. In thinking through this process concretely, we are led naturally to an *infinite regress*. The finite increment can only be estimated on the basis of the calculated values, which, however, change in the process. The numerical methods proceed in just such a way that the calculation continues step by step, the amount of the step being adjusted to the size of the steps. It is well known that, aside from approximate calculations, this method plays a special role in proofs of existence. With these, the regress is in fact not handled concretely but established according to certain criteria for convergence of the process.

In differential equations the structure of a formula appears in place of a structure based on geometry or temporal form.

## The Path from Differential Equations to Operators

A step characteristic of the new methods can be understood with the example of operators. Already in elementary mechanics, speeds and impulses are obtained by differentiation from the coordinate and energy functions. Just as the coordinates as functions of time give an exhaustive description of a movement, so can energy as a function of place and impulse (and possibly of time) present the total behavior of a mass point. The same is true for a mechanical *system* (even when generalized coordinates and impulses are introduced according to the degrees of freedom). Finally, it is also true for non-conservative systems. By way of certain differential operations one thereby obtains the relationships between instantaneous values (as differential equations) and out of that the total evolution in time. The differential equations of motion thereby result by setting a certain differential expression equal to zero.

The emergence of differential expressions can be clarified with the aid of the vibration equation. In an elementary way, it is written down for the so-called harmonic oscillator (mass point under the influence of a quasi-elastic force):

$$m \frac{d^2x}{dt^2} = kx$$

in which $m$ is mass, and $k$ is the spring constant of the oscillator.

In the form *here*

$$Lx = m\frac{d^2x}{dt^2} + kx = 0$$

*L* is a linear operator:

$$m\frac{d^2}{dt^2t} + k$$

Applied to the coordinate *x*, it supplies (by setting equal to zero) the differential equation of motion.

Emerging within this trivial example (by setting the operator equal to zero) is not the position of the mass point, nor even the concrete progress of the motion, but the virtual totality of all possible motions of the "classical" oscillator, which have *k* and *m* as parameters.

The interaction of different oscillators could be shown by operators that obey very simple mathematical rules. However, in the case of harmonic oscillators of classical physics this procedure is by no means necessary.

Now, operators like *d/dt* can be handled like *ordinary* algebraic entities, and thereby we (often) obtain meaningful new operations. This was known long before the application of operators in quantum mechanics. Thanks to David Hilbert, the tools for their strict mathematical treatment became available only in the twentieth century. Others (such as Norbert Wiener and Max Born) have added the use of operators to quantum mechanics on a firm mathematical foundation with the help of those tools. The same was achieved in a somewhat different way by John von Neumann with the help of an interpretation of the operators as transformations in the so-called Hilbert space.

The result can be expressed as follows: First, one forms the Hamilton *function* for the physical system under consideration—that is, a mathematical expression that shows energy expressed in terms of generalized coordinates and impulses (and possibly time). By substituting the differential operator

$$\frac{h}{2\pi i}\frac{\partial}{\partial q}$$

for the impulse $p$, one obtains the Hamilton *operator* out of the Hamilton function. This is now written in the form

$$H(q)\frac{h}{2\pi i}\frac{\partial}{\partial q}\psi = \lambda\psi$$

as a so-called eigenvalue equation, in which $\psi$ is a function that must satisfy certain rules regarding uniqueness and normalizable nature. As a result of these rules, solutions do not exist for all values of $\lambda$. Only a countable system for "eigenvalues" allows solutions of the kind referred to. The possible "energy states" correspond to them. The $\psi$-functions are the links between mathematical theory and observation. For the current interpretation the square of its absolute value ($\psi \cdot \psi$) yields the probabilities for finding the system in the energy states in question.

I have intentionally ignored those characteristics of the statistics by which one arrives at observable values from the operator. I have previously treated the problems connected with this. Here I want to stress the path mathematical treatment has taken in general. While in classical physics there still exists the possibility of thinking—to a certain degree—that the mathematical entities of theory are substitutes for "real quantities," there resulted in relativity theory as well as in quantum physics the necessity to let go of this very conception. The reasons for such a change vary from instance to instance. In relativity theory the structure of the spatiotemporal relations between measures presses toward the forming of an invariant that yields different spatial and temporal values depending on the reference frame. The invariant quantity itself can no longer be viewed as a quality belonging to the observed pair of events. The quantity is *made to correspond* to the pair of events and makes possible, by way of certain operations, the return to more or less observable physical quantities. At the highest level of classical dynamics such correspondences already exist. These derived concepts, however, can still be interpreted more or less arbitrarily in the "realistic" conception of the mathematics involved.

The path of thought taken toward the operators of quantum physics approximately looks like this: de Broglie with the thought that he was searching for wave-formulae whose laws could be *made to correspond* to the laws of mechanical particles turned around photon theory, which assigns to certain corpuscles an energy effect spreading in wavelike manner.*

Schrödinger thereby developed $_W$, an abstract function and found an instrument in the Hamilton function of classical physics for translating the classical formulas into the new method. With that it is admitted in principle that there are no general or rigorous methods for such translations. That is to say that the operators are found by guessing; science has yet to take seriously the admission that the essential progress has been made through mathematical intuition, and increasingly, is done this way.

## MATHEMATICAL STRUCTURE AS SUBSTITUTE FOR NAÏVE REALITY

We can draw the conclusion from previous discussions that the most significant aspect of the concepts formed in relativity theory and quantum physics lies in a fundamentally new relationship to mathematics. We will formulate this as drastically as possible: The thought constructions of the older atomism are no longer valid as a description of facts; they are valid only as auxiliary pictures that are heuristic in nature. These need not be tested for inner truth. Their only meaning consists of their role of providing points of reference for gaining appropriate operators for the factual description of new phenomena. More briefly, the supposed building blocks of the world prove to be imaginative creations. Their truth lies not in the images used but,

---

\* As de Broglie described in his Nobel Prize lecture, relativistic invariance was very important for him.

## The Phenomenology and Mathematics of the New Physics

rather, in the pronouncements made possible through them. These originate from another, higher sphere to be discussed later.

By virtue of the nature of the mathematics used here, quite a new world of ontological references opens up. We will discuss two of the ontological relations as a conclusion to this chapter. One is the principle of superposition of states; the other is the resolution of a state into component states.

### The Superposition of Quantum Mechanical States

The superposition principle originates from elementary physics and is an essential basis for the comprehension of mutually independent parts of a physical system. Speeds can be superimposed without causing a disturbance for each other; simple waves of arbitrary frequencies superimpose to produce complicated mixtures. The fundamental theorem of Fourier concerns the resolution of an arbitrary wave into "monochromatic" components. Already the elementary determination of total quantity by way of consolidating individual masses and the counting of the resulting units are expressions of elementary superposition rules. Of course, the principle is not without exceptions and is not a precondition for thinking in physics. But as far as the connections can be expressed through proportionality or—more generally—through linear equations, the superpositions are valid. Mathematically speaking, this is a consequence of linearity of equations or of differential equations. The linearity of the emerging operators and "eigenvalue" equations are significant for the quantum mechanical description of states and their development in time. From these results a far-reaching law of superposition; to begin with, it applies to the functions that correspond to the physical states. This kind of mathematics therefore demands the superposition of the physical states when it is translated into the language of physics.

The superposition principle provides the solution for one of the most difficult problems, a solution that furthers the development of quantum mechanics. According to this principle, it is possible for one and the same physical system to possess a number of mutually exclusive states, without perceivably "taking on" any of them. Here, finally, the wave and the corpuscular properties of matter can be reconciled.

What I have stated thus far more or less repeats the current concepts. The described solution is less of a conceptual nature than it is a mathematical one.

Can a philosophy of science attach meaning to the statement of the superposition principle—this principle undoubtedly being mathematically and physically correct? Here I will refer back to a matter that has been mentioned and points to a concept that I believe to be insufficiently employed. I speak of the concept of the *possible,* or virtual, state.

It was a mistake of the naïve-realistic conception of classical physics to fail to acknowledge many of their statements to be what they actually were in all truthfulness—general statements about possible events. On the one hand it was sometimes thought that statements of physics were conceptual counterparts of real things and events. On the other hand, the conceptual formulation (as discussed in my introduction) was sometimes taken to be only an inter-human convention by those who had already managed to free themselves from the naïve conception that laws are descriptions of objective relations between objective things.

The phrase "general statements about possible events" would, however, be a very modest contribution to natural philosophy without the conviction of which I spoke in the introduction—namely, that in the concepts, provided they are thought of as having content, there arises the second half of that full reality that, taken as sensory reality alone, would have to remain enigmatic.

Insofar as classical physics is valid, the different possibilities are mutually incompatible. The objects of classical physics may be viewed as determined within its framework through certain initial or boundary conditions. One may ascribe any exactitude to them that one chooses. The possibilities under consideration are arbitrary within the framework of initial conditions. One can say with Mephisto: "...in the first, we are free; in the second, slaves."

Classical physics, with its strict laws, does not point to a thorough determination of the world. Rather, it lets arise in our consciousness a part of the world's constitution as simple laws—a part that extends just as far as basic idealizations are valid. Classical physics is not overthrown by the phenomena of new physics. With sufficient insight into the natural limits of classical physics one could have welcomed the new phenomena as signs of liberation from conceptualizations that are tied too narrowly to the sense perceptible.

After the liberation has been achieved (accompanied by all kinds of labor pains), one doesn't need to deny the cognitive value of the older conceptualizations or yet be bashful about the newly won concepts. These concepts reflect the ideal structure of a stratum of the world in the same way that the classical ones had done previously. This stratum is constituted in such a way that its elements are by no means sense perceptible. The particles possess the possibility of partial realization as objects of classical thought that can, of course, possess only pseudo-sense-perceptible attributes.

The possible phenomena represent the sum of the stage props with which nonphysical beings act out sense-perceptible events. That is the inevitable conclusion to which I am led by taking seriously the conceptual work in classical as well as in modern physics. Here, however, we have also reached the frontiers where proper physical conceptualization may lead and beyond which it would leave its realm in favor of speculation. The next chapter will provide further support for the interpretation formulated above.

## RESOLUTION OF A STATE INTO COMPONENT STATES

In the section on the geometrization of mechanics, we indicated its continuation in quantum mechanics. The interpretation of operators in Hilbert space is an extremely suggestive and visualizable method of aiding understanding of relationships that defy elementary description. The element of understanding arises by virtue of the fact that one can really imagine that in a space—especially in a "linear" one—one vector, in relation to any other that does not happen to be perpendicular to it, has a projection that can be conceived of as a component in the direction of the other one.

The direction northeast is different from the direction east or north. To think of northeast as being composed of north and east in equal parts is nonsense, if one could not imagine the motion in the direction northeast as a result of the superposition of a purely northern and purely easterly one. That is the psychological reason one may view a vector also as "consisting" of the sum of its components.

These elementary images make the Hilbert space visualizable. In it we may therefore combine two statements that are mutually exclusive in the material world. In a somewhat caricatured simplification, we can describe a body that rests in the center between the two loci $A$ and $B$ as a condition that is a 50:50 mixture of being in $A$ and being in $B$.

To bring the image to a conclusion, we have to add that in quantum physics the states $A$ and $B$ are virtual and that it is nonsense to ask whether we can ascertain by simply looking where the body actually is. It lies in the nature of quantum mechanical objects that they can never be realized in the same sense as can the objects of classical physics. The statements involving projections of Hilbert vectors onto others are always preconditioned by the formulation of questions that exclusively are directed at only one or the other vector. With that, the corresponding calculus proves entirely appropriate for the description of the theater performance I mentioned. Within it this rule is valid: The description is true only from the point of view of the audience. The actors, as well as their costumes and props, do not exist for the spectator outside of the context of the play. If the spectators try to get behind the stage, they still see the same stage and look at another performance. That is, they project the vector of the operator automatically onto another reference system.

## The Phenomenology and Mathematics of the New Physics

What I have a bit fancifully described is merely the expression of the fact that we can see "through the veil of maya" only by using certain windows. Each has its own laws governing what is visible. If we want to move about beyond the curtain, we have to pay for it in that we must find completely new ideas about the beings that correspond to the symbolic vectors. This is a question of widening our consciousness. But this question is nowadays not being asked. To point to it comprises the true meaning of our discussions. With these descriptions, we are definitely at the frontiers that are being described by physics. In the final chapter we will describe in different language how the frontiers might be crossed.

*First of all, there is no neutral perception by the higher senses that represents, so to speak, equal things and force in the same way as in physical sense perceptions. To the higher sense, the objects to which it is directed present themselves, simultaneously, as active entities.*

# Chapter 10

# Physical Worldview and Spiritual Science

The elaborations of the preceding chapter are not intended to presuppose Anthroposophy in any way in the sense that it could be a support or basis for the deliberations undertaken here. For the author it serves as a background for his worldview just as any scientist is served by his convictions. As stated in the introduction, the method of presentation shall speak for itself.[31]

In the considerations directly following I will proceed differently. I will use a picture of nature and imaginations of "fundamentally knowable" spiritual realities as they were given by Anthroposophy in generally available descriptions in the early years of the twentieth century. Despite the fact that I am not addressing "believers" but turn to people who, in view of the imperceptible nature of modern physics, harbor this question: *Don't the facts of the so-called microworld reveal conditions that could be described in ways other than contemporary physics dares to do in its voluntary self-limitation?*

## The Relationship between Suprasensory Entities and Sensory Perception

In the past, humankind had no difficulty knowing and recognizing "gods, angels," and other higher entities, let alone ghosts and demons, none of which were sense perceptible. In the course of historical progress, the corresponding perception disappeared first. Then, the idea came increasingly to the fore that, for what is perceptible to the senses, the causes and the principles of explanation have to be found in the realm of sensory experience. Moreover, natural science added certain thoughts, the conditions of forces, or causes and effects, for instance, whose field of action was located entirely in the outer world.

In its youngest phase, natural science had to clean out many a prejudice arising from that, especially in physics. Determinism belongs to this, as it was conceived according to the model of celestial mechanics, and was supposed to be valid in the elementary processes as well as in the macroscopic world. To this also belongs the description of elementary building blocks with images and concepts taken from the everyday physical world. People have not yet been able to accept the consequence of the fact that entities can exist who are not sense perceptible and whose effects are not ordered by the kind of lawfulness applicable to sense-perceptible *forces* of classical physics. The cause of this, of course, is a well-founded fear of metaphysical speculation.

I am quite aware of the objection that using Rudolf Steiner's research results will be viewed either as a follower's dogma or as metaphysical speculation. Nonetheless I feel obliged to state the results of my acquaintance with such content. In addition to that, I think of the process after the manner of a working hypothesis, which seems all the more justified since the development of the science under consideration itself has removed the major obstacles to an unprejudiced examination of such hypotheses. Physics in particular has freed itself from a view of causality that is too narrow. Of course, it acknowledges now as before, a cause-and-effect relationship. Through the idea of controlling the processes, we have learned in both technology and the organic sciences how to recognize and cause the nearly energy-free transmission of complex information. The relationship between other levels of being and the field of sense perceptible events was not conceivable in a scientifically satisfying way so long as there was no model for the relationship of effects. It is not my intention to speak of such a model in order to make plausible an influence from realms beyond, behind, or above sense perceptibility. It is sufficient to point to the difficulty that blocked the path for an unprejudiced close acquaintance with the thought of other suprasensory entities. It is enough if the picture of different kinds of entities from the start shows those features that have to be obtained laboriously and enigmatically in microphysics.

Before there can be any talk at all of suprasensory beings in connection with natural-scientific questions, one must speak about how they can be thought of at all, quite apart from natural-scientific ideas.

## The Suprasensory World in the Sense of Anthroposophy

The perception of the "higher worlds" does not proceed with physical sense organs. In Rudolf Steiner's spiritual science, it is described as a phenomenon of consciousness, which can be compared most appropriately with the level of pure thinking. However, this does not mean that we are concerned with the continuation of an abstraction process that would be the natural way of pure thinking. Moreover, for progress in training toward such higher insights, soul exercises are given that are meant to harmonize to a high degree not just the thinking but also feeling and will in students. The thought is not unfamiliar to natural scientists that factual expertise in experimentation requires both an artisan's skill and an extraordinary sense of responsibility. In the field of natural science, premature conclusions and biases based on older theories have often ruined the best empirical results. The request for such harmonizing of the soul can be presented to a correspondingly higher degree if we are not experimenting with apparatus but are leading the soul of the practicing person into an arena of new experiences. We describe these conditions for a path of training especially so that it becomes generally comprehensible what we mean when speaking of higher experience. If we think of such results as being possible, we will be on our guard against frivolously claiming to set limits for natural science based on our own results. The sense in which a higher world can be discussed will be seen at once in our context. The previous statements might already have made it clear that all talk about the substance of such observations takes place through comparison and pictures.

Following this preamble, I can try to sketch some characteristic features of the suprasensory world in the language of Anthroposophy without the danger of being misunderstood. First of all, there is no neutral perception by the higher senses that represents, so to speak, equal things and forces in the same way as in physical sensory perceptions. In this higher sense, the objects to which it is directed present themselves simultaneously as active entities. The higher world does not consist of materials and forces—not even higher ones—but of living and sentient beings with whom one's perceiving consciousness enters mutual interaction directly.

Second, this reciprocal action in turn consists of stages. Details about this have to be read in the corresponding descriptions. The following remarks are inserted only so that the description can be comprehended with some degree of specificity.

The first stage of higher perception consists of lawfully ordered "images" pointing to a content that wants to speak through them without disclosing itself to the perceiver directly. On account of its image nature this stage is called "Imagination." The next stage, called "Inspiration," conveys, in the language of comparison used here, the content expressed in the images, just as writing does once we have learned to read. Still, the entities that express themselves in this language, however, remain outside the field of perception. Only the third stage *connects* the perceiver with the beings that appear in the Imagination as images, and begin to speak to the perceiver in Inspiration. The third stage is called "Intuition," in modification of an expression used in philosophy.

Third, all that is perceived in this way appears to be quite different from all that is commonly called otherworldly effects.

Not only can one find much that is communicated about details from this world to be appealing, but also can it be experienced as an essential enrichment of a worldview. Yet one will never confuse what is being described about such beings with some kind of spiritualistically colored descriptions of interactions of a spirit world with the world of our common perceptions. It is necessary to make these clear definitions because in more recent times the field of telepathic and other related phenomena has been drawn into the sphere of science, due to a disappearance of many materialistic prejudices. We certainly do not want to take up the interactions between human "consciousnesses" (telepathy) or, indeed, those between human beings and souls who have died (spiritism), nor examine specter phenomena. All this has to be mentioned if one wants to speak without compromise of the Spirit in the sense of this work.

On the other hand, the spiritual world's origin from beyond sensory perception is by no means without any relationship to nature. The life of plants, the sentient life of animals, human conscious and thinking experience—for each of these there is a corresponding possibility of perceiving them in the spiritual world, as well. In this sense, the well-known expressions *ether body* and *astral body* can be understood.

## Spiritual Entities Have Relationships That Cannot Be Derived Solely from Sensory Experience

If, as is acknowledged, the ultimate entities of modern physics are not to have sense-perceptible characteristics in their essence, and if one has the impression that they are "something" beyond their mathematical form, one must certainly be on guard against the following opinion: Well, if a suprasensory world does exist—as the anthroposophists say—then show us the connection between their beings and the phenomena that appear in our experiments and are governed by formulas. Whoever enters, even just a little, the nature of what I attempted to indicate in the previous paragraphs will be able to understand the solution—that one first has to consider the special conditions of the new experiences about which we are speaking before one can expect an interpretation, especially that of the essence of matter. Perhaps this is the most difficult area.

Indeed, the description of the higher worlds as Steiner gave it, as well as his description of what is to be attained on a regular path of training, does not follow the path from the physical world into the world bordering it, but rather the reverse. The peculiarity of spiritual beings reveals itself when one acquaints oneself with such encompassing relationships as that of the well-known description of the evolution of the world and of human beings and that of reincarnation.* This can only be touched upon in the present sketch.

However, one will always think of the description of spiritual conditions correctly by acknowledging that the beings and forces have "degrees of freedom" or "dimensions" of behavior and of their mutual relationships, which are not *directly* expressed in our sense-perceptible physical world. To give an example, we can well think of the abundance of causal relations (which are known in daily life and its scientific treatment) as a system of relations that is nearly closed. Nevertheless, the world with its interacting causes can at the same time be viewed as an expression of something else. In human relationships such a view is taken for granted. Take, for example, the transmission of information. What takes place in the links of transmission (e.g., that of a modern multiplex telegraph) is completely comprehensible physically in the sense of a close-range effect conception. The physicist is interested in *how* the apparatus functions. He may completely ignore the intention of the technician, who all the same has arranged everything for his own purposes. The true cause for the alternating current of a telephone wire lies within the consciousness of the speaker and the true effect within that of the

---

\* See Steiner, *An Outline of Esoteric Science,* chap. 5, "Knowledge of Higher Worlds—Initiation."

listener. Similarly, the extremely interesting physiochemical mechanisms in living beings become increasingly "comprehensible" for modern physiology. At the same time, these are related to etheric processes (which not only accompany life insofar as the suprasensory view is concerned, but in fact are also its essence) in the same way as writing to language or telephone electrical processes to the content of communication—that is, the meaning of other "dimensions" and "degrees of freedom."

To condense this relationship axiomatically, one can say: All physical-sensory processes have their spiritual correlates, and they who know spiritual beings are able to recognize their traces within the sensory world, but the relationship is not an isomorphism or even clearly one to one. Taking this peculiarity of the suprasensory, spiritual world for granted, the attempt shall now be made to present the problem of matter from its point of view.

Next to the world of higher entities that have found their reflection in religions, in mythology, in legends, and in folk tales, as well as in the history of humankind, there also exist "lower" assisting beings that were known to the older visionary consciousness as so-called elemental beings. Knowledge of them barely extends into our own age, communicated by means of fairy tales and other folk traditions, nowadays sometimes in a decadent way. We must not forget, while turning our attention to this subject, that they show themselves as Imagination, and therefore—dreamlike—adopt fantastic features when the higher means of knowledge have been lost. Since I have mentioned of the loss of the older view of nature for the second time, it must be mentioned that Anthroposophy particularly here wants to bring about a change and teach a new conscious experiencing and knowing as a timely expansion of scientific knowledge.

Before I proceed in the discussion of spiritual worlds and their relationships, and also their connection to the problem of matter, I must respond to an objection. How does it happen, one can ask, that natural science of today sees no reason to include the effects of extrasensory forces or beings in their considerations? For, as one will continue in the spirit of this objection, it is true that natural science in the twentieth century had to learn to reach for quite unusual concepts under the compelling force that the phenomena of the new kind exerted upon them. What then remains of our natural-scientific attitude if we leave a path that, until now, even if it was frequently accompanied by painstaking labor, led to such magnificent results? The path of faithfulness to sensory experience.

So much for the question. Those who do not put the full weight of its importance constantly before their consciousness can rightfully be accused of dilettantism, or even of irresponsibility. In all the discussions here, the requirement has to be retained that single scientific results, or facts, must be *correct* (insofar as this may be stated within the framework of the fundamental margin of error in empirical science). Likewise, the requirement must be retained that the mathematical apparatus for the treatment of these facts proves true—that its application in the required framework of exactitude of the experiments in general leads to accurate predictions.

To answer the previous question, two viewpoints have to be considered independently. One is that many of the new concepts of this century show particularly how little the elements of direct experience enter into them and how much they are suited for pointing beyond themselves. Later we will look in detail for the starting points for this among the previous descriptions. The other viewpoint is outside of science. In order to substantiate it we proceed from the statement of the fact that modern materialism—of Eastern or Western character—is completely irrefutable. Irrefutable for a very definite reason. Materialism is a *way of thinking*, not a scientific result that would be subject to control by experiment or its logical correctness. As a heritage of the past, materialism is unconscious for most thinkers. So long as one is inclined only to go along with objects of experience that are accessible to direct or indirect measurement, one will be able to call all what enters the sphere of observation a scientific phenomenon, no matter how strange in the course of its development its character may seem. Since materialism is not concerned with facts or formulas, no refutation could be brought forward that would be acceptable to the attitude described here.

However, as one can learn from the history of science, decisive developments have received their initial impulse from considerations outside of science. I do not want to recite here the long list of discoveries that have their origin in dream experiences and other inspirations of their authors. It might suffice to point to names such as Henri Poincaré, August Kekulé, and Norbert Wiener. In mathematics, observations that lead to the anticipation of the solution to a problem in any way at all are called "heuristic." Their value is undisputed. The rules, according to which mathematical proof has been strictly carried out, are firmly established within any given era, even if subsequently they change. These conditions are somewhat different in natural science and particularly in physics. Part of the rules to which the objects of legitimate research are subject is merely largely conventional, depending on the thought habits and biases of the time. Therefore the true merit of Einstein or Bohr is perceived to lie in the audacity with which they drew far-reaching conclusions out of assumptions that really fly in the face of traditional thought habits.

When we now aim at connecting the two viewpoints, we can formulate this question: Isn't it possible that the peculiar "beings" about which Erwin Schrödinger and Louis de Broglie feel compelled to speak actually have their true realm of existence elsewhere? Is it (as a heuristic assumption) too far-fetched to view the difficulties of interpreting elemental particle behavior as an expression of their true essence having several more dimensions than the world of everyday experience? The inner necessity in thinking that can here be experienced when one frees oneself from traditional form is, of course, of quite a different kind than the one with, which, for example in relativity theory, a single assumption about the speed of light drags along everything else with the aid of the mathematics applied.

Here we reach the point I already mentioned—that we *cannot* show with examples the fecundity of such heuristic considerations. For now, we must be satisfied with understanding that the cognitive gain lies in a different interpretation. That, however, is no mean thing! When Copernicus undertook his famous new interpretation of the Ptolemaic system of epicycles, an immediate improvement of astronomical predictions was expected. *The Prutenic Tables* were immediately calculated according to the Copernican theory, but after only a few decades showed basically the same insufficiencies as those of Ptolemy previously.* In spite of that, the Copernican system was accepted as a momentous achievement by an increasing number of his contemporaries. Kepler was needed to bring about radical innovations, although they were within the framework of Copernican thinking. Copernicus' deed, it is true, showed no immediate results, but the heuristic importance for all followers was immeasurable.

## The Distance between the Subsensory World of Elementary Particles and the Suprasensory World of Elemental Beings

Following this interim discussion, I want to turn to the problem of matter. In the picture of a world of consciousness and sentient spiritual beings, the material world must appear as a boundary layer. Let's take this comparison to a higher dimensional space not literally but merely qualitatively and say: As in Plato's parable of the cave, the sensory world that presents itself to us is only a projection. In contrast to the frequently used interpretation of the world, as being such a projection in some theosophical or spiritualistic way of thinking, we merely want to use this picture to understand this thought: This boundary layer can be penetrated in appropriate arrangements of experiments. Ignorant of this boundary, one then

---

\* *The Prutenic Tables* were an ephemeris published in 1551 by the astronomer and mathematician Erasmus Reinhold (1511–1553).

believes still to be dealing with a material reality and yet is really experimenting with (or in) a layer behind it.

A misunderstanding can arise here. It could appear that by leaving the material world in this way, one would get into the spiritual world. To avoid just exactly this misunderstanding, we needed the previous sketchy portrayal of the individual beings of the supra-sensory world for higher cognition in gradual training. For this, the tools of cognition for grasping these overall relations are not gradual modifications of sense representations based on perceptions. In order to use once again the comparison with a reality of more dimensions, one would like to say that its geometry is not reached by an examination of the boundary layer, be it ever so exact, but by ascent to configurations of a different kind.

Conversely, it can be understood from the overall course of evolution of mankind why at one time there should exist a stage on which the fiction that nothing else existed anywhere could be sustained with great logical consistency.

For Anthroposophy, the meaning lies in the fact that only on such a stage can beings who are endowed with a personality learn to achieve independence from their spiritual origin. Instead of inserting here a discussion about details of world evolution, the biblical phrase from the story of the temptation may be pointed to: "Ye shall be as gods, knowing good and evil" (Gen. 3:5). Not only is it a temptation—it is a promise. The similarity to God can only be attained in freedom and separation from "the gods." Who, however, brings into existence this world that seems to function with laws devoid of spirit? The gods? Yes, with the aid of helping beings that we encounter at the boundary of the material world.

The problem of matter is posed to natural scientists in such a way that, while searching for its ultimate building blocks, they encounter entities that cannot be grasped from a world that is merely mathematically comprehensible but only described statistically. For a science of the spirit the same problem exists in the fact that the beings that so to speak weave the boundary of the material world are neither the first ones that we meet in the correct progress toward knowledge of this world, nor are they beings whose actions have the same goals as the impulses "of the gods." One can continue the thought thus far expressed and, at the same time, remain faithful to the principle that higher entities cannot be *derived* in the same way as the fields or corpuscles of physics.

Thus, it is a justified to ask: Which of the material boundary phenomena calls for an interpretation? With this, the following questions emerge right away: How does in such a light the polarity substance force appear? How, in the same light, do the further distinctions matter-energy and corpuscle-wave stand? What is expressed in the "atom as phenomenon" in contrast to the "atom as hypothesis"? What is signified by the classical fields of physics that appear like proxies for perceptions? What is expressed in the fact that they have not yet been satisfactorily quantified?

Another group of questions is connected with other, less sense-perceptible relations: Can one understand the contrast between the two classes of elementary particles (fermions, bosons) better if one conceives of them as "traces" or "expressions" of entities that are not perceptible to the senses? In other words: Do the empirical rules about half or whole number spin, in connection with the applicability or inapplicability of Leibnitz's principle of the identity of indiscernibles, help in gathering these with other principles, like with the one of Pauli, into a unified vision? One would wish not a derivation but a meaningful interpretation arising out of an expansion of concepts.

The search for answers to the questions thrown open here is and was naturally felt as a summons; This is also true of "pure natural scientists." Of necessity, parts of such answers have to take on a more philosophical shape, since the quest is not merely for a new, perhaps better description of phenomena but for interpretation and meaning.

From the start I did not necessarily want to arrive at new formulas, whether they are viewed with regard to intellectual mastery or practical control of phenomena. Thus, we may be permitted to add a little to those approaches that can contribute to the struggle for new concepts that should supplement the relations already mastered mathematically. One such is the following: Beings interweaving in a counterspace can reveal themselves in physical space—but they do not have to do so. Statistics about this will be correct but say as much and as little about *these* beings as population statistics say about individual human destinies.

Two aspects of the world of these entities could be surmised from their revelations: a continuous and a discontinuous one. Both, taken in isolation, *deceive*. The one by creating for the human senses the illusion of large units of continuous expansion, the other by showing appearances naïvely conceived as body-like. As to the first, we have to consider that insight into nervous system activity teaches that human consciousness obtains knowledge of physical surroundings only by discontinuous impulses through single nerve fibers.*

The other aspect shows us in the appearance of phenomenal atomism those appearances at the point where we approach the "veil of maya," as though with a magnifying glass. The resulting single, so-called particles almost prevent the formation of a satisfactory summation into "macro-phenomena."**

This doubly illusory aspect raises the question whether the true revelation of the spiritual may be found at another boundary that is well known to all who occupy themselves with the possibilities of human consciousness. This boundary is found as opposite to the boundary of the material world. Certain differences did not yet exist for the older states of consciousness in earlier ages. To them, all of what increasingly appears to us as separating out from unity was still fused into oneness: sensory experience and the being that is

---

\* Even if we were justified in believing in a thoroughly "continuously structured external world," we would have to familiarize ourselves with the thought that its experience is transmitted to us via isolated sensory cells and discrete ganglions.

\*\* Special difficulties also appear when we attempt to consolidate the continuous "fields" according to the requirements of quantization with mathematically correct pictures.

experienced.* The abstract concept took its place. The earlier consciousness was still able to perceive beings of the spiritual world in dreamlike symbolic imaginations, referred to as mythical by contemporary psychology.**

At the present time, one can still encounter the spiritual world at this boundary when one engages upon the search in an *appropriate* way—in short, the character of knowledge presented by Anthroposophy. Anthroposophy points to *one* experience of a purely spiritual nature that reaches across this boundary into everyday consciousness of human beings. It is the only one that is familiar and accessible to every human being: the experience of one's own "I."

The "I" character of other knowledge has to be worked for. It is possible to achieve this from *thinking,* but only if the thoughts are experienced as having *content* with corresponding clarity, not only as images with ideas with limits of associations. This is at one and the same time an experience accessible to every human being and testable from one's own experience.

In the main part of this book I directed the examination to thinking as experienced content because of the convictions discussed here. I am conscious of the fact that the objection can be made that I am merely following a proposition by Steiner in this, which then might have become dogma for me. In the face of this argument, readers can be invited only to an unprejudiced examination of whether they encounter in their own mind these very questions of knowledge, in their primal form, free of prejudice. The naïve question: Properly speaking, the *what* of matter, energy, and electricity aims at *being,* and is bent only later toward other formulations through the use of scientific conventions. In the introductory chapters, we discussed the limits into which positivism wants to put the *what* of things for questions.

---

\*     The experience of the being is mostly known by hearsay, or it is experimented on by way of hallucinogenic drugs.

\*\*    In a rudimentary way, something similar does still exist when one experiences directly the essence of a person from the language of a human countenance.

## Physical Worldview and Spiritual Science

To the understanding of an encounter with spiritual beings in full knowledge belongs an important peculiarity of their appearance at this "inner" frontier of consciousness. We stated earlier that the envisioning consciousness enters into interaction with the beings that are at the same time revealed as the percepts of higher sense experience. This peculiarity consists of the fact that the *form* of imagination stems from the consciousness of the researcher. Through it the being expresses itself in the same way as meanings are expressed by language (Inspiration). Modern physicists are well acquainted in their science with the new relationship between observer and object. It is no longer possible to remain in the static attitude of "onlooker consciousness" as one could, at least in principle, in classical physics. A symptom can be found in this for what we pursued in recent observations: *Science is about to go beyond the frontier of the material world*. Accordingly, the relationship between observer and object changes. It should be noted that only after lengthy discussions on the peculiarity of suprasensory knowledge such a symptom can be interpreted.

The fact that the form in which suprasensory knowledge emerges originates in part from the observer could be taken as a motive to think of knowledge as inexact or at least doubt its scientific value or even deny the possibility of objective knowledge in this field. But experimental research knows well enough that the results of observation *also* depend on the researcher. It is part of the character of the art of experimentation to exclude such influences if at all possible and—wherever unavoidable—to remove them by means of appropriate corrections. Through this modern natural science, a training ground is provided, which finds no equal in any other area of life as far as strictness and self-criticism are concerned.

For Rudolf Steiner, the founder of Anthroposophy, the presence of natural science and its methods provided the necessary conditions for developing a scientific knowledge of the spiritual worlds. He himself moved through the study of natural science and mathematics and then occupied himself for years with Goethe's natural science, working his way through all those ideas and their "scientific value" in the sense referred to here.

Therefore, it is self-understood that spiritual cognition knows the interaction between observer and object in accordance with natural-scientific method; it is now clear that this same interrelationship challenges us with tasks as important and difficult as those meeting the experimental researcher. This circumstance is therefore far from objecting to the possibility of spiritual cognition adhering to a natural-scientific method; much to the contrary, it proves its character to be adverse to any form of nebulous mysticism.

But it must be acknowledged that the relationship between observer and object characteristic of quantum physics is of a different kind than those disturbances that experimentation generally had to deal with. Occasionally, it is said that the unavoidable disturbances caused by the observer in the area of quanta are on the order of the size of the phenomena in question and therefore, *in principle,* cannot be removed. Since Werner Heisenberg's epistemological criticism, the fundamental "uncertainty principle" has been made plausible. The discussion concerning the way in which the observer *"really"* determines the result of an experiment has subsided. Now the discussion is centered more around questions of interpretation for the applied statistical methods.

Nevertheless, most essential is the viewpoint that the apparatus that is being engaged "produces the phenomenon." The phenomenon, for whose demonstration the apparatus has been designed to begin with! In spite of this, the phenomenon is not an "artifact." One knows that it is the only way with which the apparatus can respond to the incoming activity. The statistics of impulses (e.g., the Geiger counter) contain the information about the object being studied. For this, the theory provides a mathematical form that can also be adjusted for other apparatus. Of course, this mathematical form is not the being itself but its proper expression in the theory.

Why then do these "beings" present themselves in this way? Because their appearance at the frontier of the sense-world is subject to the same law according to which the suprasensory can manifest only in human consciousness—an imagination, for example, which is not the *being* but an *expression* of it.

Objective manifestations lead to statistics as their indication for the lack of congruence between the suprasensory with the sensory at the frontier to the sensory world.

## Is the Current Path of Theoretical Physics the Only One Possible?

No doubt, important progress in theoretical physics has arisen from certain imaginative pictures. Physicists, as time passed, got used to calling these pictures models. Nevertheless, these pictures have an important feature in common with those that are called *"imaginations"* in the science of the spirit represented here. They are not intended to be images of actual sense-perceptible conditions but are free creations whose actual task consists of providing support for further knowledge.*

If we want to engage ideas like those developed in the previous paragraphs at all, for the fructification of physical research, then we are led, by the nature of the matter, to a series of questions. These address the younger generation of researchers going out in search of new ideas for the nuclear forces that are not discussed in this volume at all and for many other topical questions that are contained in Heisenberg's equation of 1958.

An immediate question is whether abandonment of the "atom as building block" must find its expression right into the mathematical description. We know that there are atomistic phenomena. We also know, however, that precisely because these formerly assumed building blocks show up manifestly in experiments that their characteristics are completely new; each particle is to be correlated with a wave and each wave with a possible particle, and neither exists by

---

\* In physics, this means to provide starting points for making mathematical formalisms concretely applicable.

itself in the sense of ordinary bodies. In the crystal lattice, in the kinetic theory of gases, in the electron gas, in semiconductor theory, and in the atom, the individual particles are necessarily taken in very different degrees of being observable.

To clarify this matter by way of an elementary example, we will use the stoichiometric formulas of chemistry. These are generally conceptualized with the help of the idea of atoms. However, this is, as is perfectly well known, by no means inevitable at the level of elementary chemistry. One does not need a model to have a law. Not even the stereochemical models *need* to be viewed as real configurations. Moreover, with exact knowledge of really large molecules, many complications of claims, fusions of networks, and other issues immediately arise that merely express, in the end, the fact that the original model was too primitive. Therefore, the model represents not the truth but is merely useful. To an immeasurably higher degree, this is true for the elementary particles previously mentioned.

We need them as "possible events" for our calculations. We should hold in our consciousness that the calculations neither can nor mean to express more than probabilities just for possible events. The elements entering them are in part as superfluous as are the internal forces that are thought to hold a system together, in contrast to energy formulations, like Lagrangian mechanics, where they need not appear. There is the difference, of course, that the inner forces in classical mechanics can, in principle, be observed.

A similar example offers itself with the so-called degenerate states of macrophysical bodies. Is it necessary to think the "freezing up of the degrees of freedom" realistically, as was common in the kinetic theory of gases? Doesn't the forming of the models up to this time show that, given the help of the basic premise of quantization, just about every mechanical system would do the same service in explaining the well-known deviations of the specific heat at lower temperatures?

Leaving this comparison behind, let's now turn to the superfluidity of helium and the superconductivity of metals. In both cases we have macrocosmic bodies whose behavior can be explained only with difficulty and with quite uncommon auxiliary ideas out of their atomic structure. It's not too difficult a prophecy to assume that more and more examples will show up involving semiconductors, as well as lasers, which allow such behavior to become the rule rather than the exception.

Today the Pauli principle regulates the behavior of separate macroscopic systems—two bodies of superfluid helium, for example, connected by a capillary—without any conceivable interaction; to avoid forming a model, they are treated as one quantum-mechanical system. By now it's a habit not to want to imagine a transmission by way of some kind of substratum. I will pick up this example directly.

### Total Gestalts in Physics

In nature-philosophical discussion of the implications of the new quantum mechanics, attention was given to the consequences for biology from its inception. A peculiar difficulty for the mechanistic point of view and its various forms in classical physics consisted of the belief that appearance of living entities as totalities in no way could be made to fit a mechanistic picture. Teleology was increasingly despised following the victory march of the doctrine of evolution. The purposeful development of the single organism from the embryonic form through the adolescent form to the mature form, the complicated metamorphoses of certain animal species, and the potential for regeneration that permits the organism to reconstruct the injured totality—all of these actually presuppose that the final form acts toward itself, so to speak, as a goal. Teleology in a narrower sense presupposes a consciousness that, knowing the goal, controls the process. The utmost that classical physics could rise to was a certain understanding for collective configurations through using external principles. These however were not viewed as either consciously or unconsciously striven for goals but as certain forms of expression of a structure of relationships effective from the start (e.g., in the differential equations that govern a process).

Physical understanding of biological processes was supposed to be achieved according to this pattern. To this day it has not been accomplished. Given this, the innovations in physics were doubly interesting. On the one hand, a possibility of thinking seemed to open up that offered escape from the determinism of classical physics. The self-regulating behavior of a growing organism and its ability to regenerate itself could not be made to fit the classical picture. One was able to hope, it is true, to penetrate sufficiently complicated structures conceptually, which would then make the *seemingly* teleological behavior in ontogenesis explainable. On the other hand, the Pauli principle represents the pattern of effects of states that act on each other, without mediation. Even if the words *total gestalt* seem somewhat pretentious for the fact that in a system such as, for example, Bohr's model of electron orbitals no two states may be occupied, the implied "knowledge" of an electron newly to be fitted in as to what conditions of energy it may not adopt is nevertheless very enigmatic. The configurations of all the other electrons form a kind of gestalt to which the newcomer has to adapt.

Some pioneers in quantum mechanics thought: Couldn't the goal of a physical explanation of biological processes be achieved according to this pattern that is unthinkable in classical physics? Altogether, thinking in physics had to get used to the following: To an event that happens here and now, what could have happened but didn't is mutually determining.

Certain considerations that have not become permanent stock in science were discussed by researchers such as Henry Margenau. Within the problem of the so-called weak interactions, two systems "know" of one another through having been in interaction once before and with a new interaction behave in a different way than two from unfamiliar worlds approaching each other. This thought leads necessarily to a discussion of the concept of information, which we will undertake in the next section.

If the pictures of atomic configurations, even in degenerate conditions thought of as being chaotic, such as at the critical point, must play a role in modern physics, then how much more important will be the elementary configurations in chemistry, and especially in biochemistry, than they are in physics, which only provide a general framework. Why then should such considerations be limited to microconditions and the so-called macrostates only be superpositions of the microstates? This we have to ask ourselves when we question whether the current path of theoretical physics is the only possible one. Thus, it is a necessary consequence of present developments to search for mathematical methods that can deal with total gestalts from the start! In the section on counterspace I again shall come back to this.

## Energy-free Transmission of Information and the Future of Physical Formulas

In his well-known and oft-quoted "universal formula," Werner Heisenberg did not intend to create a sensation or a formula that unlocks all problems as if it were a passkey. Rather, he wanted to draw the consequences from the trend of development—namely, that the numerical values of, for instance, the masses of mesons and other elementary particles only begin to make sense when they arise as different possible solutions of one and the same equation. These, of course, were supposed to be relativistically invariant and, in their structure, take account of various known symmetry laws of the particles investigated to that time. His deliberations almost inevitably led him to the "appearance" of these formulas. He says:

> A theory that correctly reflects mass and the characteristics of elementary particles from out of a simple basic equation for matter, is also at the same time a unified field theory. The circumstance that all elementary particles can change into one another was learned from experiments, and it points to the

fact that it is hardly possible to sort out a certain group of elementary particles and to find a mathematical representation only for this group. Through this experience and through the fundamental significance of symmetry characteristics, every attempt at a theory of elementary particles, as for instance the attempt made in the above equation, receives a peculiar character of inclusiveness. One can find structures that are intertwined and tied together in such a way that one can actually undertake no further changes without putting all relationships into question.

Here one is reminded of the intricate arabesques in Arabic mosques, in which such a great number of symmetries are made manifest simultaneously, that one could not change a single leaf without disturbing the relationship of the whole profoundly. And similarly, as these arabesques express the spirit of the religion from which they arose, so also the spirit of the natural-scientific epoch that was ushered in by Planck's discovery is reflected in the symmetry characteristics of quantum field theory.

However, we are standing here in the middle of a development whose results one will only be able to survey in the years to come. In that half of the century whose separate phases I attempted to portray for you, Planck's discovery led to a place from which one believes the goal can be seen in clear outline, namely the understanding of the atomic structure of matter from simple mathematical symmetry characteristics. Even if one approaches the development of the last years about which I spoke, with that brand of skepticism that belongs to the highest duties of the natural scientist, one may be permitted to say that here one has met structures of quite unusual simplicity, inclusiveness, and beauty. Such structures we deem especially important because they do not anymore concern a special area of physics, but the world as a whole. (p. 182)[10]

This (which Heisenberg describes as the remarkable completeness of an arabesque) is certainly merely a transition on the path, on which more encompassing, less symmetric, total gestalts become mathematically uncontrollable, because formulas do not have to be expressed only in spatial and temporal dimensions and in mass-energies, but also *information quantities* are met in them. These do not need to signify a measurable informational content immediately. They can serve, at the beginning at least, to express things of the kind that have, until now, been accommodated into formulas as structural elements according to Heisenberg's precedent. The future of mathematical formulas then is predictable to a certain degree while not implying that one is thereby taking the position to form them concretely.

Nevertheless, the role of this kind of information and its transference between merely virtual and not real events must not be underestimated. Experiments that show actions at a distance, following the Pauli principle, do not need to provide a direct cause for new laws and new formulas. It is quite sufficient if they permit insight into the necessity for building a new story on the building of theoretical physics. The difficulty that the classical physicist had with the understanding of the $\psi$-function in an abridged formulation can be put into the words: The temporal evolution of a $\psi$-quantity signifies nothing in the sense of physical observations and, despite this, gives in every stage the possibility—by virtue of the artifice of the formation of $\psi \cdot \psi$—of a concrete statement concerning observations, though only in the form of a probability. At the next higher stage, the informational objects to be introduced will relate to the $\psi$-functions, as these do to observations.

We can contribute an account here of *the path principally opposite* to the one outlined here. We shall formulate this in the question: What must a mathematics look like that knows total gestalts a priori and makes the extra-spatial and extra-temporal interaction conceivable, without violating justified conclusions of previous theories? For this purpose we shall outline in the following sections a view of a space, suggested by Steiner, which is already suitable today to supply the means for the shaping of the second path, for those speculations by physicists that are becoming necessary.

## Rudolf Steiner's Counterspace—A Thought Form That Still Needs to Be Made More Concrete: The Duality Principle

Intuitive space in the field of sense perception was taken as a valid and undoubted basis for all scientific treatment of the sensory world from the beginning of scientific geometry right into the nineteenth century. Euclid's geometry was able to gain, in but a few decisive axioms, so much from observation that all relations between purely geometrical phenomena could be connected to the fundamental elements in a strictly logical system. If we disregard the logical structure of geometry and draw our conclusion from the development already described, it is the Euclidean motions as measure-preserving transformations that constitute the essence of measure. Likewise, those non-Euclidean spaces already discussed rest on other more general transformations that share with the Euclidean the traits of elementary measure.

Projective geometry that does not use measurement can introduce a new definition of measure more mobile or flexible within itself. However, it is almost always presented in such a way that a certain trait is still taken unchanged from a naïve intuitive vision. This is the special role of the point as spatial element. It is true that nineteenth-century geometry knows the transition to other spatial elements—to spheres, for example. Here, however, the emphasis was not on an expansion of vision but on the group structure of the relations and on the possibility of attaining many new theorems of geometry merely by transferring elementary facts of point geometry on to geometry of spheres. Eventually, therefore, one returned to the space of intuitive vision having used new "geometry" as a formal aid, more than taking it seriously as an independent geometrical world.

The same holds true to this day for the (previously known) duality principle. This gives the simplest transition to other spatial elements. The *planes* of projective space possess such relations among themselves, having *straight lines* (as pencils of planes) and the *points* (as bundles of planes and rays), so that they correspond exactly to the known relations of incidence of points. This correspondence extends to the degree to which every proposition of ordinary geometry, which contains only statements of incidence, gives rise to an equally accurate proposition if one appropriately substitutes the terms point, line and plane, with plane, line and point. The terms *to connect* (of points, of points with lines, and of intersecting lines) and to *intersect* (of planes, of planes with lines, and of coplanar lines) must also be appropriately exchanged. In the bundle of rays and planes the metrical relations (not only those of incidence) can be dualized, which comes to expression in spherical trigonometry in the well-known "polar" formulas.

The *principle* consists of the fact that the basic structure of relations contains this symmetry that enters naturally into all derived concepts and all proofs so that the truth of this principle of transformation needs no further proof. A whole series of concrete transformations exists independently of this, through which, in the single instance, the corresponding one of the other can be gained from a construction of the "world."

The duality principle, as it has been described here as a purely geometrical fact, can be found in many variations, not only in geometry. Thus, in linear algebra a duality between N-Triples of numbers and linear equations of $n$ variables can be established. For $n=4$, for example, a reversible unequivocal relation between the points of projective geometry and the linear equations with four variables on the one hand and between the planes and quadruples on the other hand is possible.* By this, it was possible for the misunderstanding

---

\* It must only be considered that equations that appear multiplied with a real number are not considered essentially different and, likewise, that quadruples with proportional single values are understood as representing the same plane. The relation was formulated intentionally in the form of so-called plane coordinates; because of the duality principle, one would be able to assign points to the coordinate quadruples and equations to the planes, as is customary in analytical geometry.

to arise that geometry is not an independent mathematical field but simply a representation of a particularly simple case of structures in the area of linear algebra.*

Geometry based on the duality principle in which the plane as element of space was taken intuitively as fundamental has not been worked out. The abstract possibility of the translation was given, as we know. After all, during the nineteenth century, textbooks on projective geometry once in a while were written in such a way that in two columns not only the axioms but all considerations and proofs as, of course, also the basic assumptions were dually formulated and constructed to one another.

## Counterspace

Rudolf Steiner's indications toward such a reckoning (as just mentioned to be possible) rest on his *direct* imaginative insight. This distinguishes the course taken here (in the second part of the book) from the hypotheses one encounters elsewhere. That is not supposed to mean that one should on account of its origin use such suggestions in any different way than as a new possibility of thinking that is to be researched within the framework of physics with regard to its fertility. We are mentioning the imaginative origin of the idea of counterspace for another reason. One can gather from Steiner's description that it did not merely "occur" to him and that this is not a case of an expert in projective geometry "having this idea" to apply it dualistically to the usual geometry. Rather, Steiner was merely pointing to projective geometry in general, because thinking that seeks training can become more flexible and formative, being less bound by sense-perceptible forms. He gave a concrete description of that space, which in contradistinction to the physical world is not a space of centralized forces that are active between points (e.g., mass points or charged particles) but a space whose "origin"

---

\* Jean Dieudonné, *"La geometrie n'est qu'un corollaire trivial de l'algebre lineaire"* (Geometry is just a trivial corollary of linear algebra).

of coordinates lies at a periphery in the infinite distance, a space of *universal forces,* not central forces. Specifically, because of his direct imaginative means of description, Steiner's audience could not be expected to make an immediate connection to the duality principle of projective geometry. Only later, independent of each other, George Adams and Louis Locher-Ernst carried out the *geometrical* concretization.

The better one understands the duality principle as a genuine polarity—and not only as an isomorphism—the more convincing becomes the interpretation by Adams and Locher-Ernst. For that reason, both prefer speaking of *"polarity"* rather than duality when they think of fulfilling the requirements of this realistically expressed concept.*

For details I can recommend the explicit presentations that can be found with Adams[1] and Locher-Ernst.[20] To begin with we are only concerned with mastering a more mobile and encompassing geometry, using it as a tool, just as we learned to handle analytical geometry as part of our university entrance exams. In order to bring the following physical applications closer to intuitive vision, a few details of "counterspace" may here be described, as it results with inner necessity if one systematically takes as primary spatial elements not the point but the plane.

For ordinary perspective, a line "consists" of points; a plane is a set of points, as is a curved surface, a space with volume, as well as every elementary geometrical figure—all are sets of points. In counterspace, one would appropriately not focus on sets of points but on sets of planes. A one-dimensional set of planes is either a pencil of planes or a configuration called a developable surface. In differential geometry, one looks especially at the point-like character of a surface and at the similarly conceived measurement of distances in

---

\* The terms *polar, polarity, polar system,* and so on are reserved for a special class of relations in academic projective geometry. Almost never, however, is there any danger of misunderstanding if one uses the term *polar* next to *dual* in the sense intended here.

the surface. From this perspective, the surfaces that can be developed possess the same inner metric as the plane. If this perspective is completely abandoned and one considers only manifolds of planes depending only on one parameter, then the developable surface is the counterpoint to the skew curve.

A plane curve has, as its counterpoint, a set of planes, which belong to a point in the same way as the points of the plane curve belong to its plane. The dual counterpoint to the latter is therefore the cone as a surface (conceived of as unlimited on both sides, and to begin with, not as a cone of rays). Likewise, as the skew curve possesses a tangent in each point—as the limit of the line of connection of neighboring points when these come together into one—so does the developable surface possess a tangent in polarity to that as the limit position of approaching planes as their "generators"; and as the skew curve possesses osculating planes, which in turn envelop a developable surface, just so does the developable surface possess points on a cusp, which form a skew curve. Likewise, the plane curve has tangents, and its polar opposite, the plane cone, generators.

The second to last example was accidentally a self-polar figure. In elementary geometry, to any particular volume there must correspond in counterspace a particular part of the total "space of planes"; it may be called a *negative* volume. Think of a sphere as the simplest case. Now it is filled with points in normal (positive) space. Now, what does the sphere that corresponds to it in negative space look like? In answer one can look at two aspects. The points of the interior of the sphere carry no tangents. Instead, every plane by going through an interior point has an intersection with the sphere. In polarity to that, the "interior" of the negative sphere is formed by those planes that contain no tangents and for which all points have cones of contact (i.e. are points that have planes in common with the sphere).

I will add that the incorporation of these objects that are thought of counter-spatially into our visual space leads to the most various "appearances." A "negative sphere" does not need to appear as a sphere in Euclidean space.

Within visual space a "small volume" is nearly a point like a ball. In counterspace one has to speak about a small (negative) volume as being nearly a plane surface.*

It is well known, and is being cultivated within the framework of a small amount of specialist literature how the peculiarities of curves and planes have polar aspects. It is a pity that these purely geometric discussions have not been taken up at all in teaching mathematics, not even to any great degree in university courses.[20] In addition, as an auxiliary science of physics, the concepts of mathematics, like vectors, allow for interesting polar formations.

Concluding these geometrical considerations, I must point to one peculiarity. Projective space, as I have already stated, is essentially determined without a metric. However, it also contains the possibility of Euclidian as well as non-Euclidian metrics. On account of the law of duality such metrics can be introduced also into counterspace. The decisive factor of concrete interpretation by Adams and Locher lies just in the fact that they both conceptualized a counterspace strictly polar to the Euclidian space, and concretized it extensively by way of many examples. Thereby all metric concepts (distance and angle, for example) can be defined in a polar sense. It goes without saying that non-Euclidian metrics are possible and may gain importance in application.

## Possible Physical Applications

It has been known for a long time that the double nature of the basic equations of electromagnetism by Maxwell shows a kind of duality. Lucius Hanni tried in works from the beginning of this century to establish a connection with the duality in geometry. These works gained no recognition although Hanni's analogy, as H. Castelliz[4] proved, was one of considerable detail.

---

\* Think of a double-shelled hyperboloid with an extremely large opening angle of the asymptotic cone. For ordinary point-wise vision, the planes that lie "between" the two shells form the interior of such a "small" negative volume.

We can raise only questions: Is not the particle a concept of point space? Should not a plane carrying negative space characteristics correspond to it, just as much as the particle symbolizes positive-spatial characteristics in location and speed?

If one has freed oneself, as we strove for in the main section of this book, from attaching physical quantities to models in a naïve-realistic way, and if one instead takes the particles only as an adequate expression of possible observations, then the assumption that spatial-counter-spatial models do much more justice to the behavior of so-called particles than do the previous point-wise models gains some weight. One may think, for example, of the quantum jump that is an unrealizable idea—namely, a wave that, in theory, fills the whole of space would have to assume totally different characteristics the moment the particle it describes makes a quantum jump. This total transformation must occur all at once and throughout space.

We see that this idea is being stripped of its physical contradiction, in popular conception, by virtue of being thought of as a probability wave—that is, something that has no claim to physical existence. If necessary, one grants such nonphysical behavior to a thought construction.

How would it be if one used point-like and plane-like models next to one another, whose spaces interpenetrate, of course, but may only then be considered in interrelation when those physical events take place that contradict the customary (classical) mode of viewing?

Up to this point I am mostly concerned with avoiding the biased concretization of well-known formalism in point space. Already, with this, we cannot dismiss out of hand the consequences leading to new physical formulations of questions.

## Nonrelativistic Simultaneities

This final example has a relationship to the inconceivability of absolute simultaneity of distant events, which has almost become a dogma. Distant events, taking place in point space, can be extraordinarily near to each other in counterspace so that here the prohibition against signal transmission at speeds greater than that of light becomes meaningless. One only has to get accustomed to the idea that an elementary plane, which would be the polar counterpoint to a mass or charge point, possesses physical presence in relation to all its points, in the same way as we grant it without question to all planes of a point under isotropic conditions. Naturally, "anisotropic" planes are also conceivable for those theories under development, just as, depending on the circumstances, one has to conceive of a dipole not as being a simple point but as going in the direction of an anisotropic configuration.

In addition, the transmission of information, taking place instantaneously, which has to be postulated if one wants to somehow imaginatively picture the Pauli principle, suggests the availability of two interpenetrating spaces polar to one another. Whatever appears quite separate in a space can belong to one and the same configuration elsewhere. The following remark does not want to advocate the introduction of new universal constants. It is merely intended to help connect with the customary concepts of today. A sufficiently small universal constant, which joins space and counterspace, would yield the usual formulas when set equal to zero (analogous to Bohr's principle of correspondence). But, if different from zero, it would give a mathematically conceivable form to the otherwise inexplicable derivations.

## Total Gestalts

The crystal, the "degenerative star" of astrophysics, superconductors, superliquid quantities of helium, and the laser, all offer in the inorganic a picture of phenomena within which there appear total gestalts over vast molecular distances, total gestalts, whose parts are related to each other as if they knew of one another, and where there has been no evidence yet of a speed at which news spread. Lately quasars and pulsars can be added to this group.

Organisms possess other gestalts, with which the paradoxical character in the sense of physics is not obvious. The difficulty begins with the physical explanation of goal-oriented processes, if one wants to conceive of them causally. Since the early days of quantum mechanics, one was repeatedly pointed (Jordan et al.) to the possible significance of the new physical mode of thinking for the understanding of organisms. The concept of counterspace does not originate in the necessity to theoretically make conceivable the physical processes at the "lower boundary of the sensory world." Quite the contrary, it is oriented toward an upper boundary of the sensory world. The formative forces that carry the growing organism to an ordered final form, enabling it not only to sustain a stable condition homeostatically in dynamic balance but also to guide remarkable achievements or regeneration, are not objects of the sense-perceptible world. They should not be thought about according to the pattern of fields or other constructions that exist in close correlation to the observable phenomena. They are not distinguished from abstract *laws* of nature of the inorganic world by virtue of possessing a pseudo-material substratum but by acting as laws in a more flexible relationship with the processes that are guided but not dominated by them. These laws are literally *perceivable* to imagination. Thus, imagination relates to its object similar to the way formulas do to theirs—that is, to the inorganic world.

Counterspace is meant for such a world of formative forces. Why do we speak of it here at the lower boundary of the sensory world? Because the worlds into which one enters by stepping over the one, as well as the other, are not necessarily separated by chasms but are given to higher observation (imagination), complementing sense perception at the same time. Its conscious discernment is a matter of cognition.

With this we have arrived at the boundaries that are created as much by the framework of this book as by the limitations of physics, taken in its most comprehensive sense. No insurmountable boundaries of cognition are being pointed to, but other means of cognition are required. Yet we do not want to shy away from looking at the whole context that of necessity follows from crossing boundaries in modern physics. The more the conceptualizations of modern physics, as one might say, become "unimaginable," the more they call for new kinds of imaginations. So as not to form them haphazardly, and so that they not lead unaccountably into spheres the same diametrical character as are mainland and sea in our world, it might be justified to express unhesitatingly the thoughts of this last part. Those who have insight into the magnificent lack of bias with which physical researchers today are prepared to accept the most unusual phenomena, provided they appear empirically secured, and are prepared to apply the most unusual concepts, if they might contribute not to their understanding but to their conceptual mastery, have to feel duty-bound in presenting the direct relationship, which out of perception of a spiritual world can be stated, between the entities of that world and the formulas of physics.

## Homeopathic Dilutions (Potencies) and Cosmic Influences as Examples

Around 1920, Rudolf Steiner offered concrete suggestions for new experiments, which have been taken up by various anthroposophic researchers, but so much has changed in the natural sciences that some results seem less improbable or exotic than they did then [now a century ago]. In spite of that there exists a great barrier to any discussion of, for example, so-called homeopathic effects of potency. The barrier consists of the fact that Amedeo Avogadro's number seems to be an absolute limit for effects capable of being produced by substances. For instance, an argument takes the following course: In a mole of dissolved salt (e.g., silver nitrate) there are $6 \times 10^{23}$ silver atoms. If one creates a solution of 1 gram of silver nitrate in 10 cm³ water, for instance, then according to an easy calculation 1 cm³ of the "original tincture" contains $3.5 \times 10^{20}$ silver atoms. If now (as is customary in homeopathy), 1 cm³ of the solution is in successive steps diluted with 9 cm³ water (i.e., a ratio 1:10) and the resulting dilutions are designated in so-called potencies ($D_2$, $D_3$, etc.), then the $n$ dilution ($D_n$) contains $3.5 \times 10^{20-n}$ silver atoms.

Therefore, from the twenty-first decimal potency on it cannot be expected that there is even one silver atom remaining in each cm³. Thus, it is considered utterly absurd to expect any kind of provable effect of higher potencies (the 30th, for instance) regardless of the context. Physicists and chemists classify any such medical experiences as so-called placebo effects. At the outset, an objective proof is thought of as impossible whenever empirical research in this area is discussed. The results achieved are blamed on mistakes in the experiments or, at best, on unconscious influences by the "bona fide" experimenter.

Here, we are clearly dealing with aftereffects of a former, seemingly physically based, worldview. The conviction of the essentially material nature of the world is extremely deeply rooted—or more recently, a material-energetic one, including the forces between substances and other "half-material" particles, and energy as electromagnetic vibrations.

In the following sections, I will attempt a report on experiments accessible to me. Following that, I will again take up the thread of a proper discussion in the context of this book. A personal remark may precede this. When I was still of school age, I saw the experiments by Lili Kolisko for the first time. Her immaculate researcher's attitude and her efforts for absolute objectivity impressed me deeply.

During the course of my studies, confronted by the results and methods of natural science, a critical analysis of my youthful experience was unavoidable. However, since I did not want to become biased by a "powerful word of Avogadro's number," my criticism was directed primarily at the insufficient information of convincing parallel experiments. The more I was fortified by spiritual-scientific considerations about the acknowledgment of the *possibility* of such phenomena, the more seriously I urged those researchers active in such contexts to undertake controllable parallel experiments, especially, in order to test the question: Are these experiments reproducible in a way similar to other physical, chemical, physiological, etc. experiments? My concern was not an experimental *proof* but experimental questioning of nature. For it cannot be dismissed out of hand that there are phenomena at the boundary of material effects that are not any more reproducible in the same sense as, for instance, the experiments of classical mechanics. Quantum physics has found a precise form within *its* frame, by probability calculation. The boundaries of reproducibility of the single experiment can be encompassed within that form. However, before corresponding considerations can be undertaken in other areas, the experiment has to be examined.

Wilhelm Pelikan, a coworker for decades at Weleda AG Schwäbisch Gmünd, was long occupied with experimental tasks in a similar direction. He, too, saw himself led to undertaking parallel experiments and application of statistics for evaluation. A major experiment in 1962–63 was evaluated statistically by me. The results appeared in a joint article.[32] Concerning the setup and evaluation of the experiment, the result needs to be described only insofar as needed for understanding.

Potentizing was not done by hand to eliminate the influence of the experimenter. The growth of small wheat plants was used for the method of proof, just as in the first Kolisko experiments, and they germinated in the different potentized liquids (for reasons of magnitude limited to between $D_8$ and $D_{19}$). To equalize light and warmth conditions within the experimental room, the growth dishes stood upon turntables. The seeds were not preselected, in order to eliminate personal factors. (Only broken or otherwise obviously damaged grains were sorted out.) The turntable arrangement was an aggregate of three single tables, each of which was turning on its own axis, as well as around a symmetry axis common to all three. By those means, the single dishes described closely matched paths in such a way that differences in position were mostly equalized. All of these arrangements were selected based on previous experiences in order to eliminate disturbing influences, which might have simulated valid effects. The size of the small plants was measured after a few days of growth. It is of greatest importance that the single experiment was carried out sixfold from the start (on every turntable, two series $D_8$ to $D_{19}$ + water control).

The result of such a biological potency experiment therefore is a series of mean values of the growth of small wheat plants. Graphically transcribed over potency numbers as abscissas, one obtains a so-called potency curve. This curve shows a certain tendency described qualitatively early on in these examinations. The experiments by Pelikan allow one to supply the mean values with an empirically determined area of variability (mean square deviation). This latter justifies the statement that, according to customary statistical criteria, it is extraordinarily improbable to obtain the observed variances of the potency curve solely by the accidental cooperation of disturbances.*

The inserted graph shows the result in the mean of all parallel experiments of forty repetitions with the statistical variables entered. From the graph, one can deduce that the larger of the observed divergencies surpasses that multiple of the variable that corresponds to the probability of chance in the amount of 1%. According to experimental statistics such differences are considered "secure."

---

\* This is the actual meaning of so-called statistical assurance that takes place under the "zero hypothesis"—i.e., it is *assumed* that the intended influences have no effect—and the observed deviations in the different potencies are merely the product of the convergence of accidental variability.

*Physical Worldview and Spiritual Science*

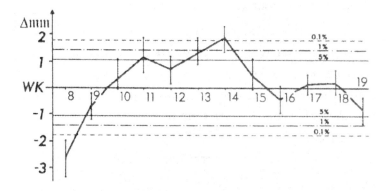

*Average Growth, entered over the potencies in comparison with the water control (WC). The arrows represent the mean square deviation calculated from forty repetitions; the horizontal lines show the significance limits for the difference against WC, independently determined by a total analysis of the variance.*

Above all, the factors here summed up show that the influences of potentized silver nitrate on plant growth can obviously be objectivized in a similar way by repetition, as is possible with other natural-scientific results. For further details and literature the publication previously mentioned[32] may be recommended.

Since exactness of measure in physics itself is about to enter similarly small quantities on the one hand, and on the other, certain concepts of cybernetics everywhere are finding entry, a few fundamental remarks might be appropriate, leading me back to the main theme. If one has freed oneself from the idea that every sense-perceptible effect can only be transmitted by material or energetic carriers, one will not immediately attempt to imagine an effective mechanism in the face of this renewed confirmation of the homeopathic thesis of the effects of higher potencies. For the sake of completeness, it may be mentioned that the modern natural scientist, when he is sure of the reality of the effects, presumably will argue as follows: When potentizing, we must be dealing with a transferable *modification of the medium*. Such a thought suggests itself for the reason that one is acquainted in biology with effect transmissions by modifications of the medium. The research represented and pursued by us wants to stay clear of constructions of effective mechanisms and especially dedicate itself to the study of what wants to come to expression in the various curves of the potencies of various substances. More recent confirmations of these types of curves reported by Kolisko, (which, for instance, are different for different metals) point to quite a new area of research.

There are further experimental methods of research that have been developed by scientists from suggestions by Steiner. They have in common that qualities that otherwise do not, or do only indirectly, express themselves as measurable quantities are made visible through the creation of *pictures* by means of sensitive processes. Especially *capillary dynamolysis* (Lili Kolisko), the copper chloride crystallization method (Ehrenfried Pfeiffer), and finally, the *drop-picture method* (Theodor Schwenk) may be mentioned here. Here is the place for a short discussion about the information concept of cybernetics. As pointed out earlier, the future of physical formulas might be influenced by the information concept. May it here be stated with a certain degree of radicality: *The more encompassing viewpoint we gain by referring to spiritual-scientific observations enables us to think forthwith in terms of information transference with carriers of equal value*—yes, even with an infinitesimal carrier in a mathematical sense. Yet it would be a fatal error if one only attempted to "explain" new realms of phenomena again with the help of a new natural-scientific concept like the one of information or *negentropy,* which places itself, according to Norbert Wiener, so to speak as a third nature force next to matter and energy.

I perceive the essentially new element of spiritual-scientific impulses by Steiner in the possibility of achieving new pictures so that new cognitive means that join mathematical formulas can lead to the kind of broadening of natural science suggested by its own evolution beyond the boundaries of the sensory world.

Briefly formulated by way of the example of potency effects: The new kinds of effects that show up here do not need to be interpreted as "information transfer," but they relate toward the materially or energetically transmitted effects *analogously,* as does the meaning of a message to the changing carrier of the transmission. We stress this here, and in spirit it is true for all methods referred to that create pictures. This is so because the cybernetic concept of information only concerns the measurable aspect, so to speak, the form alone, whereas we arrived at the confirmation that natural-scientific cognition is mainly concerned with the *content of thoughts.*

The same holds true when one is dealing with the empirical evidence of effects by cosmic influences on earthly events, whether or not such relationships are "reproducible." Recall the varying results of researchers dependent on the cycle of sun spot activity, on the phases of the moon, or of planetary aspects, or remember how long it took before the dependence of the "weather" on the moon was recognized as a scientific fact. The research is hindered by a conceptual vagueness about the point of time of a constellation. The customary confirmation of a "conjunction," for example, consists of stating the point of time when celestial "longitude" coincides. For the purpose of sky watching by the star lover, this naturally is a satisfactory determination of a conjunction of one planet with another (longitude is the magnitude of the arc, measured from the vernal point). When trajectories (orbits) are strongly inclined toward one another, great differences regarding the moment of "least distance" can already result visually. Yet this concept of conjunction, elucidated on the background of the fixed stars, does not take into account the daily movements.\* Therefore, the task of a phenomenological determination of a conjunction may be outlined here with a few words.

---

\* Here I do not include the traditional consideration of the latter, customary in astrology, since it contributes nothing to a new conceptualization of the phenomenon "conjunction."

Two heavenly bodies, like the sun and moon, determine together with the observer a plane. This meets the apparent dome of the sky in a great circle. Let's look at the movement of such a "connective great circle" of two stars! If the stars are fixed (or moving extremely slowly like the outer planets), then their connective large circle is subject to the daily movement of the sphere of the sky, that is, to a rotation around the axis of the sky. The plane of the large circle, that is the connective plane mentioned above, then envelops a cone of rotation whose axis is that of the sky. If the heavenly bodies under observation now move on the background of the fixed stars, the form of this enveloping cone goes through gradual changes. For every conjunction, there results a general cone as an envelope whose form mirrors the *conjunction as a process*.

A *phenomenology of constellations*, suggested by this thought (under strict observance of the event as it *appears* for the earthly observer), could give better preconditions for empirical formulation of questions about the effect of cosmic influences on earthly experiments. Again, just as with the discussion of potency effects, it is not the *carrier* or *medium* of such effects that is sought for, but the proof of the relationship itself. Heuristic observations about the direction in which experimental results can be looked for result without difficulty from the spatial and counter-spatial observations previously described.*

## Summary

Think back on the discussions regarding the forming of physical concepts. The observation of the development of the new

---

* The impulse for this phenomenology arose from considerations I owe mostly to conversations with P. Nanatke and W. Viersen. Wim Viersen calculated a number of narrow conjunctions between Sun and Moon at solar eclipses. There is usually such a cusp in the cone and in the line of intersection of it with the sky, but there might be as many as three cusps. The calculations can be found in the book *Konstellationen in Bewegung: Eine neue Phänomenologie von Opposition und Konjunktion*, (Constellatons in motion: A new phenomenology of opposition and conjunction, Dornach).

phenomena, as well as of the concepts formed in connection with them during this century, have led science to frontiers. These are *not* barriers to cognition that might be established by epistemological research. Also, we are not simply concerned with limits of physics, as Walter Heitler[11,12] has staked them out so strikingly. Rather these barriers prove to be, as it were in our hands, *limits of the sense-perceptible world*. We feel motivated to search for concepts that are suited for this state of affairs. That is the true intent of this book.

The physicist is mostly interested in "rational empiricism" (Goethe), in the conceptually visualizable relationship between phenomena. For long stretches in physics, idealized phenomena could be made objective. Physics of the nineteenth century, the so-called classical physics, believed that naïve-realistic pictures and causal determinism were sufficient.

The turn to the twentieth century, with its new phenomena, brought about a wrestling for new concepts. I wanted to show that it is not necessary to interpret subjectively the statistical character of the new physics in the sense of a positivist attitude. However, we are not concerned with the construction of a hidden deterministic world, as those are who defend hidden variables, and so on. We are, on the other hand, clear that one is not justified to minimize the difference between classical statistics and what is appropriate for the description of new phenomena, as is attempted by some representatives of dialectical materialism. The latter stresses the acknowledgement of an "objective, real outer world." The fundamental uncertainty of elementary processes near the limits mentioned above belongs to this concept of nature of "matter"; a more profound interpretation cannot be attempted according to this way of thinking because the "real world" (i.e., matter) is the only existing one.

We have gone back innovatively to the roots against these two basic attitudes that were born in contemporary discussions only out of the interpretation of the so-called new physics and in the drawing of ideological consequences from it. That required not only a substantial examination of the phenomenology but also our own examination of the means of cognition employed.

A different attitude at the boundary of the sense-perceptible world is permitted by the certainty resulting from the self-observation of thinking—that *reality* can be grasped in and with the content of concepts. This is what this concluding chapter is about.

After the fundamental discussion of a "suprasensory world" and its entities, and especially their relations to the boundary of the sensory world, we returned to the path of theoretical physics. The excursions into organisms and energy-free information transference, and especially into counterspace, are naturally only to be taken as examples, likewise the discussion of potency and of constellation experiments. More important than the indication given within those examples, as seriously however as they are meant, seems to be the achievement, most of all, for the correspondence of the physical worldview to be striven for, with the total concept of the world. To be sure, we consciously do *not* limit ourselves to a staking out of the boundaries of physics but attempt to orient ourselves with sober scientific attitude beyond this boundary. And we here forego a discussion of the consequences for human life of contemporary science and its technology, even though it may seem clearly called for. I leave this up to the reader, since there are many excellent publications available.

Max Jammer says: "Quantum mechanics, ultimately, is a physics of process, and not of properties, a physics of interactions and not of attributes, not even of primary qualities of matter" (p. 382).[13]

We can agree only with this characterization and reach the following conclusion: Quantum physics is concerned with processes that out of the nonphysical play into the world of matter and its forces, not with building blocks; it is a physics at the threshold of the sensory world, and its characteristic uncertainties and statistical phenomena are signs of the incomplete manifestation of entities, whose essence has to be comprehended by different cognitive means.

# Cited Works

(*listed by reference number in text*)

1. Adams, G. *Strahlende Weltgestaltung: Synthetische Geometrie in geisteswissenschaftlicher Beleuchtung Strahlende Weltgestaltung.* Dornach, Switzerland: Phil.-Anth. Verlag am Goetheanum, 1965.

2. Blokhintsev, D. I. *Grundlagen der Quantenmechanik.* Berlin: Deut. Verlag der Wiss, 1961.

3. Born, M. *Physik im Wandel meiner Zeit.* Braunschweig: Vieweg, 1957 [Eng. transl. *Physics in My Generation.* New York: Pergamon].

4. Castelastelliz, H. *Math.-Phys. Korresp.* Nr. 1–5. Dornach: Math.-Phys. Institut, 1954–1956.

5. deBroglie, L. *Licht und Materie.* Hamburg: Goverts Verlag, 1949 [Ger. trans. of *Matière et Luminaire.* Paris: Albin Miche, 1937].

6. Einstein, A. "Zur Elektrodynamik bewegter Körper." *Annalen der Physik* (4) Bd. 17, p. 811, 1905.

7. v. Freytag-Loeringhoff, B. *Logik: Ihr System und ihr Verhältnis zur Logistik.* Stuttgart: Kohlhammer-Verlag, 1957.

8. Gnedenko, B. W. *Lehrbuch der Wahrscheinlichkeitsrechnung.* Berlin: Akadamie-Verlag, 1962.

9. Gnedenko, B. W., and A. N. Kolmogorov. *Grenzverteilungen von Summen unabhängiger Zufallsgroßen.* Berlin: Akademie-Verlag, 1959 [Eng trans. *Limit Distributions for Sums of Independent Random Variables.* Cambridge, MA: Addison-Wesley, 1954].

10. Heisenberg, W. *Wandlungen in den Grundlagen der Naturwissenschaft.* Braunschweig: Vieweg, 1966.

11. Heitler, W. *Der Mensch und die naturwissenschaftliche Erkenntnis.* Braunschweig: F. Vieweg, 1966.

12. Heitler, W. *Ist ein lebender Organismus eine Maschine? Ist eine Maschine ein Organismus?* Bern: Tech. Rundschau Nr. 53. 1966.

13. Jammer, M. *The Conceptual Development of Quantum Mechanics.* New York: McGraw Hill, 1966.

14. Jeffreys, H. *Theory of Probability.* Oxford: Clarendon, 1948.

15. Jordan, P. *Atom und Weltall.* Braunschweig: F. Vieweg, 1956.

16. Jordan, P. *Anschauliche Quantentheorie: Eine Einführung in die moderne Auffassung der Quantenerscheinungen.* Berlin: Springer, 1936.

17. v. Laue, M. *Materiewellen und ihre Interferenzen.* Berlin: Akademie-Verlag, 1948.
18. v. Laue, M. *Geschichte der Physik.* Berlin: Akademie-Verlag. 1959.
19. Lindsay, R. B., and H. Margenau. *Foundations of Physics.* New York: Dover, 1957.
20. Locher-Ernst, L. *Einführung in die freie Geometrie ebener Kurven.* Basel: Birkhäuser, 1952.
21. Planck, M. "Das Prinzip der Erhaltung der Energie," *Wissenschaft und Hypothese Bd. VI.* Berlin, 1908.
22. Planck, M. *Naturwissenschaften.* 1928.
23. Poincaré, H. *World of Mathematics,* vol. 2. New York: Simon and Schuster, 1956.
24. Poincaré, H. *Wissenschaft und Hypothese.* Leipzig: Teubner, 1906.
25. Richter, H. "Zur Begründung der Wahrscheinlichkeitsrechnung." *Dialecta,* 1954.
26. Schrödinger, E. *Was 1st ein Naturgesetz? Beitrage zum naturwissenschaftlichen Weltbild.* Wien: Oldenburg, 1962.
27. Stäckel, W., and J. Bolyai. *Urkunden zur Geschichte der nichteuklidischen Geometrie.* Leipzig, 1913.
28. Steiner, R. *Philosophie der Freiheit: Grundzüge einer Modernen Weltanschauung.* Dornach: Rudolf Steiner Verlag, 1949 [Quotations taken from the Eng. trans.: *Intuitive Thinking as a Spiritual Path: A Philosophy of Freedom.* Hudson, NY: Anthroposophic Press, 1995].
29. Steiner, R. *Goethes Naturwissenschaftliche Schriften Bd. XVII.* Stuttgart: Joseph Kurschner, 1883–1897 [reprinted Dornach: Rudolf Steiner Verlag, 1974].
30. Steiner, R. *Wie erlangt* man *Erkenntnisse der hoheren Welten.* Dornach: Phil.-Anth. Verlag am Goethe anum, 1961 [Eng. trans. *How to Know Higher Worlds: A Modern Path of Initiation.* Hudson, NY: Anthroposphic Press, 1994].
31. Unger, G. *Physik am Scheideweg.* Stuttgart: Freies Geistesleben, 1962.
32. Unger, G., and W. Pelikan. "Die Wirkung potenzierter Substanzen." Dornach: *Phil.-Anth.* Verlag am Goetheanum, 1965.
33. Vester, F. *Neuere kybernetische Aspekte zur Entstehung des Lebens, Vitalstoffe Zivilisationskrankheiten.* Goldach, 1961.
34. von Weizsacker, C.F. *Die Tragweite der Wissenschaft,* vol. 1. Stuttgart, 1964.

# Index of Names

Adams, George  179, 181
Archimedes  95, 141
Avogadro, Amedeo  185

Balmer, Johann Jacob  51, 127
Becquerel, Henri  52
Bernoulli, Jacob  90–91
Blokhintsev, Dmitri  65, 66, 67, 70
Bohr, Niels  10, 16, 46, 56, 66, 68, 71, 127, 128, 130, 136, 172
Bolzano, Bernard  22
Borel, Émile  91
Born, Max  56, 58, 144
Bruno, Giordano  xv

Cantor, Georg  22, 23, 27, 28
Castelliz, H.  182
Cauchy, Augustin-Louis  142
Chebyshev, Pafnuty  90
Cherenkov, Pavel  113
Christoffel, Elwin Bruno  116
Copernicus  162
Coulomb, Charles August  38
Coulomb's law  39
Crookes, William  49

d'Alembert, Jean le Rond  115
Davisson, Clinton  71
de Broglie, Louis  8, 9, 10, 27, 70, 71, 72, 74, 132, 146, 161
Dieudonné, Jean  178
Dirac, Paul  112, 136
Dunoyer de Segonzac, Louis  52

Ehrenhaft, Felix  53

Einstein, Albert  52, 65, 99, 101, 102, 103, 104, 110, 115, 116, 124
Euclid  176
Exner, Sigmund  14

Faraday, Michael  38, 39, 55, 74
Fermi, Enrico  136
FitzGerald, George  52, 57, 99
Fizeau, Hippolyte  57, 98
Fresnel, Augustin-Jean  98
Friedrich, Walter  52

Galileo  xv, 2, 11, 141
Gauss, Carl Friedrich  100, 140
Germer, Lester  71
Gilbert, William  37
Gnedenko, Boris  90, 92
Goethe, Johann Wolfgang von  193
Goldstein, Eugen  52

Hamilton, William Rowan  76, 125, 139, 144, 145
Hanni, Lucius  181, 182
Heisenberg, Werner  xii, 29, 46, 68, 133, 168, 169, 173, 174, 175
Heitler, Walter  27, 130, 193
Hertz, Heinrich  51, 52, 135
Hilbert, David  144, 149, 150

Jacobi, Carl Gustav Jacob  76, 125
Jammer, Max  47, 194
Jeffreys, Harold  80
Jordan, Pascual  xi, 104, 141

Kant, Immanuel  40, 100

Kekulé, August 161
Kepler, Johannes 162
Knipping, Paul 52
Kolisko, Lili 186, 187, 190
Kolmogorov, Andrey 92
Kurlbaum, F. 126

Lagrange, Joseph-Louis 76
Laplace, Pierre-Simon 9, 93
Leibnitz, Gottfried Wilhelm 164
Lenard, Phillip 52
Lindsay, Robert 2, 3, 4, 81, 111
Lobachevsky, Nikolai 100, 109
Locher-Ernst, Louis 179, 181
Lorentz, Hendrik 51, 57, 58, 98, 99, 101, 103, 110, 127
Lummer, Otto Richard 126
Margenau, Henry 2, 3, 4, 81, 111, 172
Maupertuis, Pierre Louis 140
Maxwell, James Clerk 55, 98, 103, 112, 135
Michelson, Albert 99
Mie, Gustav xii
Minkowski, Hermann 113

Newton, Isaac 38, 111, 138, 141

Pauli, Wolfgang Ernst 136, 171
Pelikan, Wilhelm 187
Pfeiffer, Ehrenfried 190
Planck, Max 41, 51, 52, 53, 57, 58, 69, 73, 125, 126, 127, 136
Plato 11, 162
Plücker, Julius 49
Poincaré, Henri 161
Poisson, Siméon Denis 38
Pringsheim, Ernst 126

Rayleigh, John William Strutt, 3rd Baron 126
Richter, Hans 81, 82, 83
Riemann, Bernhard 100, 116
Ritz, Walther 70
Rubens, H. 126
Rydberg, Johannes 70
Rydberg–Ritz combination principle 70

Schopenhauer, Arthur 31
Schrödinger, Erwin 13, 14, 15, 26, 58, 122, 136, 146, 161
Schwenk, Theodor 190
Steiner, Rudolf 35, 39, 100, 133, 154, 155, 166, 167, 175, 178, 185, 190
Süssmilch, Johann 93

Torricelli, Evangelista 2

von Freytag Liiringhoff, Bruno 12
von Guericke, Otto 38
von Laue, Max 52, 53, 56, 57
von Mises, Richard 82, 84
von Neumann, John 144
von Weizsäcker, Carl Friedrich 2

Weichert, Emil 52
Weierstrass, Karl 142
Wiener, Norbert 144, 161, 190
Wien, Wilhelm 52
Wilson, Charles T. R. 52
Wilson cloud chamber 52

Yukawa, Hideki 136

Zeeman effect 52